Please
EXPLAIN

Dr Karl Kruszelnicki

Illustrated by Adam Yazxhi

HarperCollins*Publishers*

HarperCollins*Publishers*

First published in Australia in 2007
by HarperCollins*Publishers* Australia Pty Limited
ABN 36 009 913 517
www.harpercollins.com.au

Copyright © Karl S. Kruszelnicki Pty Ltd 2007
Illustrations and cover design copyright © MAXCO creative 2007

The right of Karl Kruszelnicki and Adam Yazxhi to be identified as the moral rights
author and illustrator of this work has been asserted by them in accordance with
the *Copyright Amendment (Moral Rights) Act 2000*.

HarperCollins*Publishers*
25 Ryde Road, Pymble, Sydney, NSW 2073, Australia
31 View Road, Glenfield, Auckland 10, New Zealand
77–85 Fulham Palace Road, London W6 8JB, United Kingdom
2 Bloor Street East, 20th floor, Toronto, Ontario M4W 1A8, Canada
10 East 53rd Street, New York, NY 10022, USA

National Library of Australia Cataloguing-in-Publication data:
Kruszelnicki, Karl, 1948– .
  Please explain.
  Bibliography.
  ISBN: 978 0 7322 8535 7 (pbk.).
  1. Science – Popular works. I. Title.
500

Cover photography: Jo Duck
Cover design and internal illustrations: Adam Yazxhi / MAXCO creative
Additional illustrations: 'Man' on p 200 by Max Yazxhi (aged 4^1/$_2$ years); 'DNA' on pp 21
  and 117 by Zac Yazxhi (aged 1^1/$_2$ years); Data chart on p 204 by Lol a Kruszelnicki
Internal design and layout by Judi Rowe, Agave Creative Group
Printed and bound in Australia by Griffin Press
79gsm Bulky Paperback used by HarperCollins*Publishers* is a natural, recyclable product made
from wood grown in a combination of sustainable plantation and regrowth forests. It also
contains up to a 20% portion of recycled fibre. The manufacturing processes conform to the
environmental regulations in Tasmania, the place of manufacture.

6 5 4 3 2 1   07 08 09 10

This book is dedicated to my mother and my father,
who gave me life, love, libido and the licence to be myself.

Contents

Celery with Negative Kilojoules

My teenage daughter and her friends were recently talking about 'negative kilojoules'. They were curious about whether eating celery would make you lighter — supposedly because digesting celery requires more energy than is actually present in the celery. (As an aside, my teenage son would never have given negative kilojoules a thought in his whole life. Is this a generic boy–girl difference?) So I had to check out this 'negative kilojoules' theory.

Even though I trained as a medical doctor, the lectures on Nutrition came to a grand total of only eight hours over the whole course. And because nutrition is a huge topic, I knew that that eight hours wouldn't cut it. So I consulted the experts, the nutritionists who spend years studying this enormous body of knowledge. They told me that no-one had ever done an experiment where people were fed nothing but celery. This is partly because the volunteers would soon become ill on a diet totally lacking in proteins or fats and partly because celery has so few kilojoules that any measurements would be difficult to calculate.

Negative kilojoules

Celery is about 95% water,
2.2% digestible carbohydrates and
1.8% indigestible carbohydrates,
as well as insignificant amounts of
protein and fat.

BUT there are a small number of
kilojoules found in celery ... not many ...
but there are some!

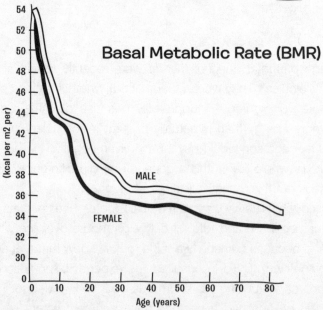

The BMR is a measure of the amount of energy needed to keep you alive
when you are resting, in a fasted state, lightly clothed and in an environment
with a comfortable temperature.

Basal Metabolic Rate – 101

One way to get to the bottom of this 'negative kilojoules' story is to look at the Basal Metabolic Rate (BMR) — a measure of the amount of energy needed to keep you alive when you are resting, in a fasted state, lightly clothed, and in an environment with a comfortable temperature.

Technically, the Basal Metabolic Rate is a measurement of how many millilitres of oxygen you breathe in for every gram of body weight per hour. At rest, human beings typically breathe in about 0.3 ml of oxygen per gram of body weight per hour. Nutritionists measure the BMR by putting a volunteer in an airtight box and measuring how much oxygen they burn up and how much carbon dioxide they generate. These changes in oxygen and carbon dioxide are easy to measure when you consume a huge meal of steak and lots of potatoes — and almost impossible to measure when you have a small meal of a few sticks of celery.

About 70% of your total daily energy expenditure is spent in keeping your organs running (e.g. liver, skeletal muscle, brain, kidneys and heart) — the BMR, while another 10% comes as a side effect of burning up your food. The remaining 20% comes from physical activity. You can see how, if you increase your exercise just a little and maintain the same food intake, you can lose weight. For example, walking uses about 2000 kJ per hour.

Basal Metabolic Rate – 102

However most of us don't think in terms of O_2ml/gm/hr — we think in kilojoules.

The 'average' man uses about 11 000 kJ per day, while women use around 8000. Of course, this will vary according to your weight, age, gender, coffee habit, exercise program, genes that your parents gave you, height, surface area, muscle mass — and yes, how much you eat.

Back in 1918, J. Arthur Harris and Francis G. Benedict published their paper 'A Biometric Study of Human Basal Metabolism'. They measured the BMRs for 136 men and 103 women and then did a 'partial correlation coefficient analysis' for any factors that might have an effect on BMR — a big task in the days before electronic calculators and computers. The equations — the original century-old ones for the total heat produced in 24 hours — are still a reasonable approximation:

Men, H = 66.4730 + (13.7516 x W) + (5.0033 x S) − (6.7550 x A),
Women, H = 655.0955 + (9.5634 x W) + (1.8496 x S) − (4.6756 x A),

where H is the total amount of heat produced/energy burnt in 24 hours in kilocalories, W = weight in kilograms, S = height in centimetres, and A = age in years. To convert kilocalories to kilojoules, multiply by 4.2. (According to current wisdom in the book trade, each equation in a book reduces the sales by 50%. That's why I've cleverly hidden these equations deep inside the chapter and bunched them up so that you can skim right over them and ignore them.)

Food Tax Overview

Surprisingly, just to digest an average meal, human beings increase their BMR by 50% — but only for about four hours. In other words, it costs energy to digest the meal. (This might be the basis of the 'negative energy in celery' myth.) Nutritionists call this extra energy the 'Thermic Effect' — a kind of 'Energy Tax'. It is also called the SDA (Specific Dynamic Action) — but whatever you call it, it takes energy to digest food.

The question is — how much of the energy in the food is needed to digest it? To answer this, you need to know what your body does to get the energy out of the food that you eat.

Food Tax Breakdown

If you have ever eaten a tough steak, you might think that a lot of the energy that it takes to get the nutrition out of your food comes from chewing. But no, chewing the food takes only 1–2% of the total energy in the food. The vast majority of the extra energy needed to digest the food is burnt in the gut.

Ultra-marathon Humans

The sport called the 'ultra-marathon' can last up to six days. And the athletes taking part might not beat the Snake Kings of BMR for the highest BMR title, but they do amazing feats. An article in the journal *Nature* reported that the athletes '... Trek 125 kilometres, and cycle 250 more. Kayak 131, rappel through canyons for another 97, and swim 13 in churning whitewater. Throw in some horseback riding and rock climbing; spread it all over six days in the blistering Utah heat; and never stop to sleep.'

One of the biggest of these events, the 'Primal Quest Expedition Adventure Race', has teams of four athletes, racing across some 700 km of very difficult terrain. The winning team gets US$250 000 in prize money.

For some unknown reason, these athletes reach their ultra-marathon peak in their mid-30s – amazingly late in life for such a demanding sport. One theory says (with good evidence) that it helps a lot to get the 'right' genes from your parents. Another theory says (also with good evidence) that the brain of a good ultra-marathoner is heavily involved, for example, in coordinating all the different biochemistries from the overworked lungs, legs and other parts of the body.

In one day, the athletes burn up 84 000 kJ of energy – about 10 times the 'resting' BMR of an average athlete. It's very difficult to eat this much.

About half of this energy is used in three stages. Some of the energy activates the muscles of your gut to mash up your food and push it along the 10-m length of your gut. This helps break down the food into small blobs ranging in size from a fingertip to the head of a matchstick. More of the energy is used to make special chemicals (enzymes) which further break the food down from little blobs into individual molecules. And in the third stage, more energy is used to pump these tiny molecules across cell membranes and into the cells that form the lining of your gut.

The other half of the digestive energy is used inside the cells lining your gut. This energy is used to stitch together the very small molecules into bigger and more useful molecules. For example, individual amino acids get joined to make proteins.

You might ask — why break the molecules down, only to build them up again? Because if only extremely small chemicals can get across the lining of your gut, bigger things (like bacteria and viruses) will find it really hard to invade. All of this activity — chewing food, breaking chemicals down and building them up again — temporarily increases your BMR from 1 to 1.5.

Diversion – Kings of Metabolic Rate

If human beings exercise really hard, for example in an Olympic 100-m sprint, they can crank up their total energy expenditure to 20 times normal — but only for 10 seconds or so. Horses, however, can do much better. In the Melbourne Cup, thoroughbreds can reach a BMR of 45 times normal for a few minutes.

Snakes seem to be the Kings of Metabolic Rate — because of their enormous meals. Human beings might eat a meal equal to 1% of their body weight, but a snake can readily eat 25% of its body weight. When the Sidewinder Rattlesnake (*Crotalus cereastes*) does this, it can increase its Metabolic Rate up to 17 times normal. The Metabolic Rate reaches this peak about 24 hours after the snake has started to eat, and stays above seven times normal for about three days. No human beings could do this.

There is a report of a Burmese Python (*Python molurus*) reaching a Metabolic Rate of 45 times its BMR after eating a meal equal to its own body weight.

Snakes are also the Kings of Gluttony. It seems that the 'record' for a snakes's meal is about 1.6 times its own weight!

High Energy Burning

Another animal that burns up energy at a massive rate is a frog attracting its mate by calling. A good set of calls gives frogs a better chance of getting a mate, so they have to go way beyond the occasional weak croak.

Calling is so important in one particular frog species that they compress their whole chest to force a massive blast of air through their vocal cords. These chest muscles are huge – 20% of a frog's total weight. And the muscles are special as well. They are packed with blood vessels and mitochondria, so that they can keep on contracting powerfully for a long time. When these frogs call for their mate, they are burning up energy at 20 times their resting BMR.

As far as staying power is concerned, these frogs are between human beings and pythons. They can keep calling three times a second for seven hours.

Kings of BMR Rock (Gutwise)

Pythons live on a razor-thin edge between too much food and starvation, so they spend most of their time waiting. They have to wait. Mathematically inclined biologists have worked out the sad statistics for a python that might cruise around looking for food. They calculated that the extra food that the snake found would not compensate for the extra energy that it would have to burn up in hunting the food. In other words, a python that cruises around hunting for food would starve to death. For this reason the python has adapted a lifestyle of sitting and waiting for weeks, and sometimes months, without eating a meal.

Pythons can really throttle back how much energy they use. When your standard python is not eating and just living off its energy stores, it will shrink its gut to stop the normal flow of digestive juices. After all, it takes lots of energy to make these juices. The python will also shrink its liver, kidney and even its heart, as starvation stares it in the face. So the python will stay completely still for days, weeks and sometimes up to six months waiting for dinner to come within range. And, of course, once our python swallows a few rats, its liver, heart and kidney will amazingly swell by up to 50% within a day or so.

Save Energy by Being Blind

Scientists have found incredible economies of energy in the animal kingdom. For example, Brian McNab, a biologist from the University of Florida, has been studying the Texas Blind Salamander. He thinks that he might have worked out why it is blind. The Texas Blind Salamander lives in underground caves where there is no light, and where the food supply is just small random scraps carried through by underground streams. Any animal that survives down there will have to conserve all its energy. And it turns out that in the eye, the retina and cornea burn up a huge amount of energy, the cells always being replaced by new cells.

At some stage in the past, there was almost certainly a random mutation which resulted in a blind salamander. Because there's no light in these caves anyway, the first blind salamander wouldn't have been at a disadvantage. In fact, being blind would have saved energy. Presumably, this mutation gradually multiplied over the generations until, eventually, all the salamanders in this cave were blind.

Back to Food Tax – The Wrap-up

In pythons, the 'Energy Tax/Thermic Effect' can reach about 30% — so the snake gets only 70% of the energy in the meal.

In human beings, the Thermic Effect is much less, perhaps because metabolic processes can be more efficient if they are regular (e.g. three meals per day). Interestingly, the energy needed to process a meal varies with the type of food consumed. This Energy Tax is about 25% for proteins; 2–3% for fats; and about 5–10% for average carbohydrates. So if you ate 100 kJ of each, after digestion you'd be left with 75 kJ from the protein, 97–98 kJ from the fat and 90–95 kJ from the carbohydrates. In general, nutritionists calculate that a reasonable Energy Tax for the average person's diet is about 10%.

But note that the Energy Tax is never greater than 100%, which is what the 'celery negative kilojoule' myth claims.

The Energy Tax on fat is God's little joke. Not only does fat have the highest number of kilojoules per gram of any food, it also has the lowest Energy Tax. This guarantees that eating fat will really stack on the blubber. Surprisingly, the Atkins diet (high fat, high protein and low carbohydrate) helps people lose weight — but only for a short time. This is because protein is the most 'filling' of all nutrients. Second, foods such as chocolate and ice cream are forbidden because they contain carbohydrates. But people can't stand this diet for more than a few months. This is probably good because a long-term diet of lots of protein would overload the kidneys.

Celery – The Bottom Line

Nibbling on a stick of celery is nothing like chewing on a chunk of cow. Celery is about 95% water, 2.2% digestible carbohydrates, 1.8% indigestible carbohydrates, as well as insignificant amounts of protein and fat. The digestible carbohydrates in 100 g of celery will provide you with 32 kJ, while the indigestible carbohydrates will give you 16 kJ (because bacteria in your bowel ferment them to make energy). And the energy needed to process all of these carbohydrates will be about 4 kJ, leaving you with about 44 kJ — a small number, but definitely not a negative number. So you do get a few kilojoules from eating celery.

One thing is certain, however — if your mouth is full of celery, there's no room to stuff in fatty, high-kilojoule foods like chips and hamburgers.

Is there a food that takes more energy to digest than there is energy in the food? Probably ice. You definitely burn up energy when you heat the ice into liquid water, and I am pretty sure that there are zero kilojoules in water. Black coffee, tea and diet cola also have no kilojoules — but they do contain caffeine, which will rev up your BMR by about 10%. However there is no research to show that caffeine addicts are any thinner than the rest of us …

Less Than Zero?

In the USA, various companies are now marketing drinks that they claim can help you lose weight. This is very interesting, considering that a study presented at Experimental Biology's 2005 annual meeting found that soft drinks account for 9% of calories in the average American's diet. (A decade earlier, white bread was the culprit.)

How can drinks make you lose weight? By including stimulants that increase your BMR. Currently, flavour of the month is epigallocatechin gallate, an antioxidant found in green tea and caffeine.

It all sounds boringly familiar. Heroin was introduced to combat addiction to morphine, valium was the 'perfectly safe' drug to help you sleep, and amphetamines would help you lose weight. Have these companies never heard of 'side effects'? Better Living Through Chemistry is not always the way.

Fibre and Carbohydrates – Confusing Definitions

'Fibre' is a confusing term, when it refers to food that you eat. Fibre, according to nutritionists, is definitely not bamboo stalks or the stuffing inside old mattresses. It's an old term, used to refer to 'indigestible carbohydrates', which are in reality partly digestible. (Sorry about that – life can be confusing.)

'Carbohydrates' is a term that dates back to the early 18th century, when chemists studied glucose ($C_6H_{12}O_6$) and other similar chemicals. They noticed that they could write their chemical formulae as 'hydrated carbon' – a bunch of carbon atoms (C) joined to a bunch of water molecules. (Glucose would therefore be seen as six C atoms plus six H_2O molecules.) Nowadays, we know that some carbohydrates don't follow this simple formula, but we're stuck with the name.

'Digestible carbohydrates' are your regular carbohydrates. The various enzymes in your small intestine will break them down to smaller chemicals, releasing about 16 kJ of energy per gram.

'Indigestible carbohydrates', the so-called fibre, should really be called 'partly digestible carbohydrates', but that's a big mouthful. They can't be digested in your small intestine. However the bacteria in your large intestine can partly break them down and get some energy out of them. It's a win-win situation – both you and the bacteria get energy out of the 'fibre'. In fact, if you didn't have the bacteria there, you would get no energy out of the fibre – but because of the bacteria, you get about 8 kJ per gram.

Energy Tax in Birds

Human beings are pretty 'lazy' when compared to other critters. For example, some tiny Antarctic birds can maintain a high Metabolic Rate for days, even weeks.

While in Antarctica, Mark Chappell, a biologist from the University of California at Riverside, studied cute little Adelie Penguin (Pygoscelis adeliae) chicks. These baby penguins appeared to be totally lazy – begging food from their parents while sitting perfectly still in the same spot for days at a time. But Chappell discovered that they were actually working very hard.

If the stomach of an Adelie Penguin chick happens to be empty, the bird is burning up energy three times faster than human beings do. And when they get fed, they double this rate of burning energy again.

Why do these baby penguins burn up so much energy? Probably because they are the preferred food of a predatory seabird called the skua (or jaeger). These predatory birds see the cute penguin chicks as small, juicy and defenceless food targets. So the baby penguins need to bulk up as quickly as possible so that they will appear more menacing and look as if they could defend themselves against an attacking skua.

Any sumo wrestler or couch potato can tell you that if you really want to put on some weight, you should eat a lot and do nothing. This is why the penguin chicks just sit in one spot and grow fat and, hopefully, big enough to avoid getting eaten by a predatory seabird.

Big and Little Calories – Confusing

I would like to apologise for the fact that calories can be big or little.

The standard definition for 'calorie' in Physics books states that one calorie is the amount of energy needed to heat one gram of water by one degree Celsius. But nutritionists define their 'calorie' as being 1000 times bigger than the Physics 'calorie' – and they use the same spelling for the word 'calorie' and don't even have the Big Calorie spelt with a capital 'C'.

Until I realised this, it confused the heck out of me.

References

Harris, J. Arthur and Benedict, Francis G., 'A Biometric Study of Human Basal Metabolism', *Proceedings of the National Academy of Sciences*, December 1918, pp 370–373.

Pearson, Helen, 'Freaks of nature?', *Nature*, 21/28 December 2006, pp 1000, 1001.

Roan, Shari, 'Less than zero', *Los Angeles Times*, 27 November 2006.

Secor, Stephen M. and Diamond, Jared, 'Adaptive responses to feeding', *The Journal of Experimental Biology*, 1995, pp 1313–1325.

Wolke, Robert L., *What Einstein Told His Cook*, New York: W. W. Norton & Company, 2002, pp 6, 7, 25–27.

Zaidan, Frederic III and Beaupre, Steven J., 'Effects of body mass, meal size, fast length, and temperature on specific dynamic action in the timber rattlesnake', *Physical and Biochemical Zoology*, April 2003, pp 447–458.

Near-Death Experience - The Living End

Most of us have a reasonable idea of what a Near-Death Experience (NDE) is. It's supposed to happen while you are in the actual process of dying — but if you don't actually die, you get to talk about the NDE.

During an NDE you might hear strange noises and have feelings of blissful peace and joy sweeping over you. You might then float out of your body, seeing it from above, and head towards a tunnel. You enter the tunnel. At the far end of the tunnel is a bright light. As you get close to the light, you are met by someone already dead, who tells you that this is not yet your time, and regretfully, you are plunged back to our prosaic planet with the rest of the Great Unwashed. In fact, this rather complex hallucination comes from your own Personal Theatre of the Mind.

In a nutshell, it is commonly believed that the NDE happens only to those who are dying and that it is proof of an afterlife — but neither belief is correct.

History of NDE

In 1926, Sir William Barrett, a Fellow of the Royal Society, published a slim volume on deathbed visions, in which the survivors described to him the basic elements of the NDE.

Stay away from the light

A Near-Death Experience (NDE) is supposed to happen while you are in the actual process of dying – but if you don't actually die, you get to talk about your NDE.

However, the NDE became very popular in 1975, with the publication of Raymond Moody's bestselling book, *Life after Life*, which recounts the anecdotes of over 50 survivors. At first, there was some resistance to the concept of the NDE, but with further research, it was found to be surprisingly common.

Over the years, researchers have gathered some statistics.

About 60% of people have a positive NDE experience and about 30% are separated from their body. About 23% enter a dark tunnel, about 16% see a light in the tunnel, while 10% actually enter it.

Who Has an NDE?

About 10–20% of survivors of heart attacks report having an NDE. Similar figures come from survivors of electrocution, coma, attempted suicide, near-drowning, massive shock due to blood loss, bacterial infection and anaphylactic shock.

As you can see, NDEs can happen under many circumstances. And there doesn't seem to be any common factor.

NDE Without Death?

You don't have to be dying to have an NDE.

Near-Death Experiences can happen even when you are not close to dying (despite the 'D' in NDE). For example, the training of combat fighter pilots can involve high G-forces in a centrifuge, which forces the blood away from the brain to the feet — leading to unconsciousness. Dr James A. Whittery, former chief aeromedical scientist for the US Naval Air Warfare Center, gathered data on some 1200 pilots, more than 200 of whom experienced NDEs. The typical pathway involves a greying of the vision, followed by total blackout and vivid dreams. Some 20% of pilots who blackout in this centrifuge training later report classic NDEs with the out-of-body experience.

And to further separate the NDE from death, look at hospital situations where patients report an NDE. Half of them would have survived without any medical intervention — in other words, they were nowhere near death. And yet they had classical NDEs.

Consciousness

We still don't know exactly what causes an NDE, simply because the brain is so fiendishly complicated — as is the whole concept of consciousness. Consciousness begins with knowledge about the world both outside *and* inside our skin.

The brain receives information from our primary senses, i.e. vision, sound, taste, smell and touch. Skin sensors add information on temperature, pain and pressure, while sensors in the ear provide information about balance and orientation. Other sensors in the joints, tendons, muscles and bones tell the brain where our limbs are located.

All of these bits of information, and many more as well, flow into our brain like many rivers into a sea, to give us the strange perception that we call 'consciousness'.

Cause of NDE

As a result, there are dozens of theories about what causes an NDE. These include bizarre electrical activity in parts of the brain; endorphins in the brain stem; as-yet-undiscovered hypothetical chemicals in the brain such as endopsychosin; intrusion of our normal REM dream sleep into our consciousness and stimulation of the vagal nerve. All of these theories have some elements of fact to back them up and are probably all involved in varying amounts, depending on the person having the NDE.

But recently, Dr Olaf Blanke, a neurologist at the École Polytechnique Fédérale de Lausanne in Switzerland, was able to consistently duplicate some of the elements of the NDE in the laboratory. He stimulated, with electricity, a part of the brain called the left angular gyrus. Patients immediately reported the impression of a shadowy person lurking behind them, parallel to them. And when he stimulated the right angular gyrus, they reported that they left their physical body and could look down on it from above — the classic out-of-body experience.

Living End

At some stage in the future, we might know enough about the brain to fully understand the Near-Death Experience. But right now, we know two facts about the NDE. First, you don't have to be dying to have one. Second, the NDE neither proves nor disproves the existence of an afterlife.

OBE = Out-of-Body Experience

In an Out-of-Body Experience (OBE), a person sees their body and the world from a location outside their regular physical body. They feel that they are awake, even while looking down at their sleeping body.

These experiences are surprisingly common. If you look at people who have had at least one OBE in their whole life, the statistics range from 8% in Iceland to over 50% in 'special' groups such as marijuana users. The experience usually lasts for less than a minute.

References

Blackmore, Susan, 'Near-death experiences: in or out of the body?', *Skeptical Inquirer*, 1991, Vol 16, pp 34–45.

Blakeslee, Sandra, 'Out-of-body experience? Your brain is to blame', *The New York Times*, 3 October 2006.

Blanke, Olaf, et al., 'Induction of an illusory shadow person', *Nature*, 21 September 2006, p 287.

Fox, Douglas, 'Light at the end of the tunnel', *New Scientist*, 14 October 2006, pp 48–51.

Nelson, Kevin R., et al., 'Does the arousal system contribute to near death experience?', *Neurology*, April 2004, pp 1003–1009.

van Lommel, Pim, et al., 'Near-death experience in survivors of cardiac arrest: a prospective study in the Netherlands', *The Lancet*, 15 December 2001, pp 2039–2045.

Leprosy, Limbs and Lies

I was recently rereading a favourite article, 'A short history of leprosy in postage stamps', when I suddenly wondered what the average person thought leprosy was. I asked a random passer-by (okay, my daughter) and she said something along the lines of 'their hands and legs fall off, and sometimes even their heads'. I'm sure that she meant their noses rather than their heads. However, like many of us, she believed that leprosy caused bits of your body to fall off — but this is not true.

History of Leprosy

The word 'leprosy' comes from the ancient Greek words *lepros* meaning 'a scale' and *lepein* meaning 'to peel'.

Leprosy has afflicted human beings for many centuries. It existed in ancient China, India and Egypt as far back as 6000 BC.

It was probably introduced into Europe by Alexander the Great around the 4th century BC. It later reappeared in Europe with the returning legions of Pompey into Italy after the war with King Mithridates of Pontus in the 1st century BC. There is evidence of leprosy in Great Britain and France as far back as 500 AD. The Vikings then apparently carried it back to Scandinavia from Great Britain. (By 1230 AD there were 250 leper hospitals, or leprosaria, in England alone.) The incidence of leprosy in Europe peaked in the

14th century. At this stage, there were about 300 leprosaria across northern Europe. To warn uninfected people to keep their distance lepers had to carry a warning bell. However, they were allowed to gain comfort from their religion. For this reason small, round 'leper holes' were made in the walls of many churches, so that they could at least see a church service — but from the outside.

Although already on the decline in Europe, leprosy was brought to the New World by explorers in the 16th century. It was later introduced to Australia and the Pacific Islands. In fact, between 1920 and 1929, 35% of the entire population of Nauru was infected with this disease.

In Western Australia during the 1930s, Aboriginal lepers were confined above the 25th parallel, which became known as the 'Leper Line'. They were sometimes kept in chains while waiting for a ship to transport them to the Commonwealth Leper Station in Darwin. Once they arrived, they were crammed into very small compounds or shelters, often left open to the weather. However, white patients with leprosy were sent to the far more salubrious Wooroloo Sanitorium

Leprosy and Stamps

Many countries have released stamps relating to leprosy.

In 1948, Cuba celebrated the International Leprosy Conference when the disease was first given its modern classification. In 1964, Belgium released a stamp for World Leprosy Day. In 1973, the centenary of Hansen's discovery of the bacterium, more than a dozen countries commemorated his work with stamps.

In 1971, Fiji released a stamp when it closed its old leper hospital. This is the first and probably the only stamp relating to leprosy that has been released by any British administration.

(an infectious diseases hospital) near Perth. The white lepers lived in small comfortable rooms (with fireplaces) near the sea, where the food was excellent and the staff cleaned for them.

In the past lepers were expected to live in isolated leper colonies. Such colonies still exist today in India, the Philippines and Vietnam.

Leprosy rates in Australia are only about one case per million, with higher rates in indigenous Australians and immigrants from countries where the disease still exists. Worldwide, leprosy is endemic in Angola, Brazil, Central African Republic, Democratic Republic of Congo, India, Madagascar, Mozambique, Nepal and the United Republic of Tanzania. There are also instances of leprosy in parts of the USA.

Leprosy is now known as Hansen's Disease

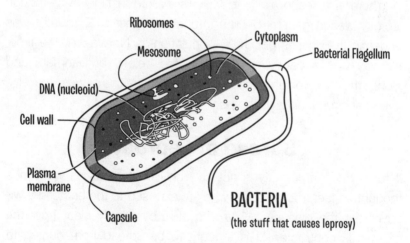

BACTERIA
(the stuff that causes leprosy)

Leprosy is now called Hansen's Disease after Armauer Hansen, the Norwegian who in 1873 discovered the bacterium (Mycoacterium leprae) that caused Leprosy.

Christianity and Leprosy

St Francis of Assisi (1181–1226 AD) took special interest in caring for lepers. Around the late 1200s, many Christian Knights formed various Orders to care for the unwell, including lepers. One of the first of these was the Order of the Hospital of St John of Jerusalem.

In 1873, a Belgian priest called Father Damien set up a leper colony on the Hawaiian island of Molokai. He contracted the disease in 1884 and died of it in 1889.

Cause of Leprosy

Leprosy is now called Hansen's Disease, named after Armauer Hansen, the Norwegian who in 1873 discovered *Mycobacterium leprae*, the bacterium that caused it. In fact, it was the very first bacterium to be discovered that caused a human disease.

There is a reason why it was discovered in Norway. Although leprosy was in general decline throughout most of Europe, it was endemic (two cases per 1000 population) in Norway. At the time, Norway was a deprived and neglected colony of Denmark and along the Norwegian west coast in winter there was massive overcrowding, thanks to the poor housing conditions.

Spread of Leprosy

It turns out that leprosy is not as highly contagious as was once thought. In fact, most people who are exposed to the bacterium will never get the disease. While it is still not exactly clear how the disease is contracted, it is thought to be spread from human to human by respiratory droplets. It is probably not spread through intact skin but possibly through broken skin.

After exposure it can take a long time to develop leprosy, even up to 50 years. The average incubation period is about 5–10 years.

Pathology of Leprosy – Overview

Leprosy affects the skin, the nerves and the lining of the upper respiratory tract.

Think about it as a two-stage disease. First, the bacteria attack you and, in return, your immune system responds. Second, you lose the motor and sensation functions of your nerves.

It then gets more complicated. Leprosy affects people differently, depending on their genetic make-up.

In fact, genetic make-up is so important to leprosy that 90–95% people who are exposed to the disease never become infected. Only 5–10% of people are susceptible to leprosy. Some of those saintly missionary types were susceptible and some were not — but they didn't know it at the time.

If left untreated, leprosy can cause collapse of the nose, clawed hands, dropped feet and bending and shortening of the fingers and toes. It can also cause blindness.

Pathology of Leprosy – Details

If your immune system responds vigorously, you will suffer a mild version of the disease — the paucibacillary (or tuberculoid) type.

With this type of leprosy your skin is less affected. Because your own cells crowd in around the invading bacteria in an attempt to seal them off, small nodules (or tubercules) appear in the skin. The robust reaction of the immune system involves the full thickness of the skin around the nodule, including the sweat glands and the nerves. This results in a firm, dry spot on the skin where there is no sensation of temperature, and very importantly for the myth, no sensation of touch. The immune system response is a double-edged sword. It stops the bacteria from spreading rapidly but also causes damage to the local tissues. Usually, the disease progresses slowly and irreversibly, although this type of leprosy will sometimes spontaneously disappear as your immune system overcomes it.

However, if your immune system responds weakly, you will contract the multibacillary (or lepromatous, or cutaneous) version — and will be much more severely affected by the disease.

In this case, the skin is badly affected. The bacteria are not walled off in your tissues. They spread widely, travelling along the nerves, but do not stick to them (as they do if your immune system responds strongly). In the face, the bacteria make the skin thicker and corrugated, giving you what are called 'leonine facies'. The nose is overloaded with bacteria, which can destroy the wall between the two nostrils, causing the nose to collapse.

The Myth

Leprosy can attack skin and nerves, causing areas of skin to lose both pigment and sensation. This is the key point about the myth.

The nerve damage results in loss of sensation and motor function. It is this lack of sensation that results in the most tissue damage. If you don't have feeling in your hands or feet, you are more likely to injure them — and then neither notice nor treat the damage. It is this frequent trauma that results from lost sensation that causes the tissue damage, not the bacterium *Mycobacterium leprae* rotting the flesh.

Leprosy *does not* rot flesh.

Treatment

Treatment for leprosy first became available with the drug dapsone in the 1940s. However, the bacteria soon became resistant to it, making it virtually useless by the 1960s. Today's multi-drug regimes are effective, and are used to try to avoid the problem of drug resistance. Prolonged courses of treatment are needed because leprosy is a chronic infection. Unfortunately, delivery of antibiotics is still a problem in many poor countries where leprosy is a public health issue.

In 1991, the World Health Assembly stated its aim of reducing the numbers of people who needed treatment for leprosy to fewer than one per 10 000 population — about 600 000 people in the world. But today the number of people with leprosy is somewhere between two and 15 million people.

For the World Health Organisation to achieve their goal in reducing these figures, it is our duty to overcome the stigma associated with the disease and its sufferers — and for us to stop treating people with Hansen's Disease like lepers.

Two or 15 Million?

The range of people with leprosy seems awfully wide – from two to 15 million. How come?

First, leprosy can take 50 years to develop. This means that many people infected with the bacterium do not have any symptoms yet.

Second, when the symptoms do appear, they are initially nonspecific. The symptoms could be indicative of many diseases, one of which is leprosy. Therefore these people are not diagnosed with the disease either.

Third, because of the social stigma and public revulsion directed at people with leprosy, sufferers will successfully hide the disease for many years. In the past, people with leprosy were not just feared, ostracised and shunned – they were also shot, flogged to death, buried alive, poisoned and burnt at the stake.

References

Gordon, Briscoe, 'Disease, Health and Healing: Aspects of Indigenous Health in Western Australia and Queensland, 1900-1940', 'Chapter 6: The Great Fear', A thesis submitted for the degree of Doctor of Philosophy of The Australian National University, September 1996.

Navon, Liora, 'Beggars, metaphors, and stigma: a missing link in the social history of leprosy', *The Society for the Social History of Medicine*, 1998, Vol 11, No 1, pp 89–105.

Rinaldi, Andrea, 'The global campaign to eliminate leprosy', *PLoS (Public Library of Science) Medicine*, December 2005, pp 1222–1225.

Wood, S.R., 'A short history of leprosy in postage stamps', *Proceedings of the Royal Society of Medicine*, August 1974, Vol 67, pp 717–719.

Where Did the Day Go?

There has long been a spurious claim that a NASA computer 'proved' that a miracle from the Bible actually happened — the so-called 'Missing Day of Joshua' or 'Long Day of Joshua'.

Science and religion are often seen as being on opposite sides of the fence, i.e. one is based on 'proven facts and data' while the other is based on 'ancient fables, faith and belief'. However, occasionally, the gap is said to be breached, as in the case where a computer supposedly supported the biblical miracle of the 'Missing Day of Joshua'.

I first saw this particular, supposed collusion of science and religion in the mid-1970s — as a badly typed chain letter in those days. Since then, it has resurfaced in widespread emails, religious internet forums, newslist groups and from the mouths of preachers. I have even seen it engraved on wooden plaques hanging on the wall in people's homes. And it is still sent to me via snail mail or email about once every fortnight.

The Miracle

The specific miracle referred to is in *Joshua* 10: 12–13. Fighting the Forces of Evil, Joshua sets off from Gilgal, defeats the enemy (with some divine help) and returns safely.

Josh. 10:9 After an all-night march from Gilgal, Joshua took them by surprise.

Josh. 10:10 The Lord threw them into confusion before Israel, who defeated them in a great victory at Gibeon. Israel pursued them along the road going up to Beth Horon and cut them down all the way to Azekah and Makkedah.

Josh. 10:11 As they fled before Israel on the road down from Beth Horon to Azekah, the Lord hurled large hailstones down on them from the sky, and more of them died from the hailstones than were killed by the swords of the Israelites.

Josh. 10:12 On the day the Lord gave the Amorites over to Israel, Joshua said to the Lord in the presence of Israel: 'O Sun, stand still over Gibeon, O Moon, over the Valley of Aijalon.'

Josh. 10:13 So the Sun stood still, and the Moon stopped, till the nation avenged itself on its enemies, as it is written in the Book of Jashar. The Sun stopped in the middle of the sky and delayed going down about a full day.

Josh. 10:14 There has never been a day like it before or since, a day when the Lord listened to a man. Surely the Lord was fighting for Israel!

Josh. 10:15 Then Joshua returned with all Israel to the camp at Gilgal.

Although Joshua seemed to be winning on the battlefield, he called for divine help. The Lord answered Joshua's pleas, prolonging the sunlight by 'about a full day' so that Joshua had enough light to win the battle. This is the famous 'missing day'.

Joshua

Joshua was appointed leader of the Israelite tribes after Moses died. A brilliant, courageous and charismatic warrior and leader, he overwhelmed Canaan and distributed its lands among the 12 tribes of Israel.

The book of *Joshua*, the sixth book of the Old Testament, tells his story in three main parts. The first part relates how he conquered Canaan, the second covers his distribution of the Canaan lands to the Israelite tribes and the last part is about his farewell speech and death.

The case of the missing day

Monday

Tuesday

Wednesday

Thursday

???

Friday

Saturday

Sunday

A religious miracle ... scientific fact ... or simply a 'sleight of hand' ... where has the missing day gone?

The Scientists

What's so special about this miracle?

Well, the story goes that this miracle can resolve the ongoing problem that some NASA scientists still have, even today. These particular scientists work at the Greenbelt Goddard Space Center in Maryland. Their computer calculates what's called the 'ephemeris' — a catalogue of what goes on in the heavens. It's a list of the locations of the Sun, Moon, planets, stars and satellites etc., with respect to any given reference point. They need this information as a reliable navigation aid to guide their various spacecraft to their destinations. They usually run their computers only forward in time, to make sure that their spacecraft will go where it's intended.

The story continues that on one occasion, purely as a test exercise, the NASA scientists ran a computer simulation back in time, covering thousands of years. It ground to a halt with a pathetic blaze of flashing red lights, having come across an unbreakable logic error. According to the story, the computer had found 'The Missing Day'.

This was supposed scientific proof of the miracle in which God had stopped the world turning so that Joshua could better slay his enemies.

History of Myth – Part 1

This myth began back in 1890 when Charles Adiel Lewis Totten published his book, *Joshua's Long Day and the Dial of Ahaz — A Scientific Vindication*. Charles Totten was not a scientist of standing but an Army lieutenant who had been palmed off to Yale University from 1889 to 1892 to teach military tactics and science. An anti-Semite, he was obsessed with Ayrian racial purity and the imminent apocalypse. He 'showed' (using very dodgy methodology) how Joshua's miracle neatly explained a one-day gap in the supposedly 6000-year age of the Universe. (You can see one problem right away — the Universe is many billions of years older than 6000 years.)

In 1927, a certain US Creationist, Dr Harry Rimmer, published his book, *Harmony of Science and Scripture*. He devoted an entire chapter, 'Modern Science and the Long Day of Joshua', to how Lieutenant Totten converted an agnostic astronomer who had come across this missing day, by showing him Joshua's exploits in the Bible.

Surprisingly, Totten did not even write this story — it was a total fabrication by Rimmer.

History of Myth – Part 2

Now computers enter the story.

In the 1960s, Harold Hill, who claimed that he was a consultant to NASA, began telling this same tale about Joshua's miracle and the missing day. Around this time, NASA was very much in the public eye because of the race to the Moon, so claiming a link to NASA was very hip. Hill suitably modernised his fable by adding some newfangled impressive technology called 'computers'. One of his talks was taped, transcribed and then passed on to the journalist, Mary Kathryn Bryan. She then published the story in her regular *Mary Kay's Kollum* in the *Evening World* newspaper (in Spencer, Indiana) on 10 October 1969, a few months after the first landing on the Moon. Although the story attracted incredible interest at the time, it then faded away only to find new life on the Internet.

But there are still problems with the truth of the story.

First, various newspaper journalists tracked down Harold Hill and interviewed him face to face. He was never able to provide them with any documentation confirming that the computer had found the missing day.

Second, NASA at Greenbelt released a statement denying the incident. They wrote in a press release that they had 'no knowledge of the use of its computers supposedly by Mr Harold Hill, and attributed to our scientists. Goddard does not apply its computers to the task of projecting thousands of years into the future or the past, as this would be irrelevant to the operational lifetime of satellites, which rarely exceeds a dozen years'.

Third, NASA also denied that Hill had even been a consultant for them. They wrote that Hill 'worked briefly at Goddard early in the

1960s as a plant engineer, a position which would not place him in direct contact with our computer facilities or teams engaged in orbital computations'.

The Answer

Dr Rimmer lied in 1927 as did Harold Hill in the 1960s. Hill was never a 'consultant' to NASA, but President of the Curtis Engine Company in Maryland. His company maintained diesel engines for NASA.

But ignore the lies and exaggerations and just think about the concept of a missing day.

You can measure the length of a piece of string only if you can reach both ends of the string. And you can tell if it's missing a bit of length only if you can measure between both ends of the string.

In the same way, you can find a missing day in time only if you have known dates on each side of the missing day. We have good calendars today, *after* the time of Joshua, but what about *before* his time?

Eclipses are ideal for this purpose, because they are such well-documented and memorable events. In the 21st century, we now have records of eclipses both after and before the time of Joshua. But at the time of Harold Hill's lectures, the earliest documented eclipse was in 1217 BC, nearly two centuries after Joshua battled the forces of evil. In Harold Hill's day there were no eclipses documented before the time of Joshua and, therefore, there was no way to find a missing day. It is mathematically impossible.

Bert Thompson, an American Creationist and religious man himself, wrote an article on this missing day in the publication *Reason and Revelation — A Monthly Journal on Christian Evidences*. He wrote: 'We do a disservice to God's Word when we attempt to "defend" it with stories like these that with a bit of common sense and a small amount of research, can be shown to have no factual foundation whatsoever.'

There is no point in telling lies for Jesus.

More Day Stopping

There is another example in the Bible of The Lord interfering with the length of the day – in 2 *Kings* 20: 9-11. King Hezekiah had been promised that his illness would be cured. He was doubtful and asked God to give him a sign.

9 And Isaiah said, This sign shalt thou have of the Lord, that the Lord will do the thing that he hath spoken: shall the shadow go forward ten degrees, or go back ten degrees?

10 And Hezekiah answered, It is a light thing for the shadow to go down ten degrees: nay, but let the shadow return backward ten degrees.

11 And Isaiah the prophet cried unto the Lord: and he brought the shadow ten degrees backward, by which it had gone down in the dial of Ahaz.

References

Brunvand, Jan Harold, *The Truth Never Stands in the Way of a Good Story*, 'Chapter 10: The Missing Day', Chicago: University of Illinois Press, 2000, pp 137–148.

Thompson, Bert, 'Has NASA discovered a "missing day"?', *Reason and Revelation: A Monthly Journal of Christian Evidence*, May 1991, pp 17–19.

Ovarian Cancer and Pap Smears

About 50% of Australian women over the age of 50 believe that a pap smear will detect ovarian cancer. This is a very dangerous and potentially fatal myth.

Surprisingly, the survey to find what women thought pap smears would identify was carried out by the National Breast Cancer Centre — but oddly, they're the ones managing the 2007 National Ovarian Cancer Program. (Yup, that's three different cancers to think of — ovary, cervix and breast.)

Cervical Cancer

A pap smear can pick up early changes in the cells of the cervix. If left untreated, these changes can develop into cancer.

The cervix is the bottom part of the uterus and sits in the top of the vagina. Because of its location a doctor can take a sample from the cervix relatively easily when women attend for their regular pap smear test. Although I say 'relatively easy', there is some discomfort involved.

In Australia, the Pap Smear Program has been very successful at reducing the rate of cervical cancer in screened women. With regular pap smear testing cell changes are discovered and treated early, hopefully before any abnormal cells develop into cancer. Even so, there are still some women who are not regularly screened and so miss out on the benefits that come with early detection.

At this stage, I have to mention Mrs Papanicolau, wife of Dr Papanicolau. He did a pap smear on his wife every day for 30 years, and so gave us a baseline for the normal cells of the cervix. Luckily our current pap smear screening guidelines do not recommend a pap smear as often as this. Just one smear every two years for eligible women, who have always had normal smears in the past, is adequate.

Symptoms of Ovarian Cancer

The 2007 National Ovarian Cancer Program survey found that over half of the women questioned could not name a single symptom of ovarian cancer.

The symptoms can include abdominal bloating, unexplained weight gain or weight loss, changes in bladder and bowel habits, abdominal or pelvic pain, indigestion and unusual fatigue.

As you can see, these symptoms are very nonspecific and can occur for any number of reasons. In fact, in some women these symptoms can be part of their 'normal' life. As a result, ovarian cancer is usually diagnosed quite late in the illness.

Women with new, unexplained symptoms that persist should consult their local doctor. Perhaps nothing will be found, but it is very important to have any symptoms checked.

Ovarian Cancer

Although the ovaries are quite separate from the cervix, they are also located in the lower abdomen. These almond-shaped organs sit on either side of the body of the uterus, and are usually about 3 cm wide. Their job is to produce eggs and secrete hormones.

Ovarian cancer is more common in older women, usually occurring in women over 50 years of age. The most common ovarian cancer is epithelial ovarian cancer which originates in the covering of the ovaries.

The Ovaries and Pap Smears

Pap smears **DO NOT** detect ovarian cancer. Pap smears pick up early changes in the cells of the cervix. If these changes are left untreated, they could develop into cancer.

The Pap Smear test

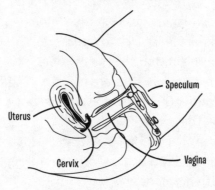

The cervix is at the bottom part of the uterus and sits in the top of the vagina.

The 'most amazing' Uterus

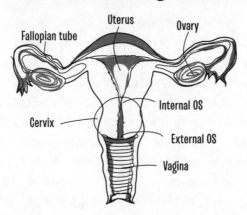

The ovaries are quite separate from the cervix and they are also lower in the abdomen. They sit on either side of the body of the uterus, are almond-shaped and usually about 3 cm wide.

Types of Ovarian Cancer

There are four main types of ovarian cancer, each with its own natural history, treatment and outcome.

- Epithelian ovarian cancers make up 90% of all ovarian cancers. In turn, there are five different subtypes of epithelial ovarian cancer – serous, mucinous, endometrioid, clear cell and undifferentiated. They all come from the surface of the ovaries.
- Germ cell ovarian cancers are rare, tending to happen in women under 30 years of age. They arise from the eggs inside the ovary.
- Sex-cord stromal ovarian cancers are uncommon and can happen at any age. They arise from the tissues that secrete hormones.
- Borderline ovarian cancers are actually epithelial cancers, that are not as aggressive as the other ovarian cancers.

There is currently no general population screening test available for ovarian cancer. (However, women who do have genetic family cancer genes should be reviewed annually, and have ultrasounds, blood tests and physical check-ups and examinations.) The lack of a reliable screening test for the general population is another reason why women tend to find out that they have ovarian cancer only when the disease is well advanced. A later diagnosis affects how well or how badly women fare with treatment. Studies done between 1992 and 1997 showed that only 42% of women diagnosed with ovarian cancer were still alive five years later. On the other hand, if the ovarian cancer is both diagnosed and treated early, over 80% of women will survive more than five years — the basic yardstick for cancer patients' survival.

Genetics of Ovarian Cancer

First, 95% of all ovarian cancers happen in women without genetic risk factors. Second, many women with these risk factors never develop ovarian cancer.

About 5% of ovarian cancers are related to inherited genetic factors. Most hereditary ovarian cancer is related to mutations in the BRCA1 gene. Fewer hereditary cancers are related to mutations in the BRCA2 gene.

HNPCC (Hereditary NonPolyposis Colorectal Cancer) syndrome is related to an increased incidence of ovarian cancers and other cancers of the gut and the genitourinary system.

Women who have a first degree relative (mother, sister or aunt) with ovarian cancer, and no other family history of this disease, have a risk of ovarian cancer only slightly above that for the general population. The vast majority (97%) of these women will never develop ovarian cancer.

Women who have symptoms suggestive of ovarian cancer are usually investigated by a combination of a physical examination, a blood test for a tumour marker, and an ultrasound or CT scan of the pelvis. Surgery is necessary for a definite diagnosis and the treatment of the cancer. This is best done in a medical centre that specialises in treating ovarian cancer, as the disease is usually widespread. Chemotherapy is most often given to help destroy the cancer.

One good side effect of the oral contraceptive pill is that it offers some protection from ovarian cancer, as does pregnancy and breast-feeding, which have the effect of 'resting' the ovaries.

Ovarian Cancer Stats

Ovarian cancer is the fifth most common cancer in women. It is less common than cervical cancer. About one in 70 women has a chance of developing ovarian cancer during her life – a lifetime risk of 1.4%. Most cases (90%) happen in women older than 40, the mean age at diagnosis being 63 years.

During the reproductive years of her life, a woman will pop out some 400 eggs from her ovaries. We do not know exactly why, but it is protective with regard to ovarian cancer if the ovaries pop out fewer eggs. For example, during the nine months of a typical pregnancy, the ovaries will pop out no eggs at all. So risk factors include having few or no children, starting periods early and finishing them late in life and having the first child after the age of 30.

Take Home Messages

There are three Take Home Messages.

First, new or persisting symptoms need to be checked, no matter how vague.

Second, there is no screening test for ovarian cancer at the moment.

Third, a pap smear, effective as it may be, looks only for early signs of cervical cancer, not ovarian cancer.

References

Neesham, Deborah, 'Ovarian cancer screening', *Australian Family Physician*, March 2007, pp 126–128.

'Symptoms and ovarian cancer', Editorial, *The New York Times*, 19 June 2007.

It's All a Bit of a Yawn

During a meeting at the end of a long and hard day spent working on a totally new project (in the daring field of comedy-science TV), a bout of contagious yawning suddenly spread through the group. The person who started this mini yawning epidemic said, in her defence, that she must be low in oxygen. This was why she had yawned so widely — it was merely a silent, natural scream for more air (and oxygen). Like all good myths, it has a tiny whiff of truth.

Yawning, O_2 and CO_2

Certainly, after a few hours of constant breathing at rest the lungs become 'stiffer' and less compliant. And yes, one decent yawn will return the lungs to their usual flexibility. However, this does not affect blood oxygen levels — at least not in healthy people.

One of the leading yawn experts on the planet is Robert R. Provine, Professor of Psychology at the University of Maryland, who has been studying yawning for decades. In 1987, his classic experiment found the link between yawning and the amount of oxygen that the volunteers breathed in. Although normal air is mostly nitrogen (about 80%) and only about 20% oxygen, for this experiment he ditched the wimpy 20% oxygen rubbish. He went all the way and gave his volunteers 100% oxygen to breathe, but they still yawned just as much. The five-times increase in oxygen had no effect.

Professor Provine also found that most people never yawn when burning up lots of oxygen, e.g. when exercising hard.

He also tested whether high levels of CO_2 made you yawn. Carbon dioxide is manufactured in your body as a waste product. You breathe it out to get rid of it. The normal CO_2 level in the atmosphere is about 0.03%. He gave his subjects levels 100 times greater — 3–5% — and it didn't make them yawn more often.

So you can forget about low oxygen or high carbon dioxide being the causes of yawning.

In the Beginning ...

The word 'yawn' comes from the Middle English word 'yanen', which in turn came from the Old English word 'geonian', which means 'to gape' or 'to open wide'. So yes, chasms really do yawn beneath you. A yawn is a long, slow and deep inward breath, taken with a wide-open mouth, followed by a short outward breath lasting for an average of six seconds. The eyes sometimes close at the peak of the yawn, and water a bit.

About half of all stretches will be accompanied by a yawn. However, it doesn't really work the other way. Only about one in nine yawns are accompanied by a long, slow and luxurious stretch of the arms and legs. (The fancy word for a simultaneous 'yawn and stretch' is 'pandiculation').

Practically all animals with backbones yawn. In most species, the males do most of the yawning, but in human beings, the balance is about 50:50 for men and women.

We start yawning even before we are born, about 11 weeks after conception. It seems to help protect the baby's developing lungs. The baby floats in amniotic fluid, a mixture of the baby's urine and a liquid made in the lungs. The baby's lungs do expand and contract but 'breathe' amniotic fluid, not air.

We don't know why unborn babies yawn. Perhaps yawning moves stagnant fluid out of the lungs or removes webs of tissue that might block the airways.

Parents of young babies have told me that contagious yawning is a 'one-way street', the baby's yawns will make the parents yawn

[Cut to man yawning]

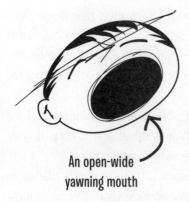

An open-wide
yawning mouth

The word 'yawn' comes from the Middle English word 'yanen', which in turn came from the Old English word 'geonian', which means to 'gape' or 'to open wide'.

A yawn is a long and slow deep inward breath, taken with a wide-open mouth and followed by a short outward breath lasting for an average of 6 seconds.

but not the other way round. Parents have also informed me that they are able to 'spread' their yawn to the baby at around 8–36 months of age. I guess that the baby is Top Dog in the house, at least for a while.

Of course, yawning is very important when you travel on aeroplanes. Yawning opens up the eustachian tubes, which run from the middle ear to the back of the throat. A few good yawns will relieve the pain of the air pressure, when the plane is taking off or landing.

Weird Yawning Links

Yawning has a number of strange links to different aspects of human experience.

For example, the yawn has a strange link to sexuality. In male rats, medications that make them yawn and stretch will also give them an erection. In most human beings, antidepressant drugs such as clomipramine and fluoxetine can put a damper on sexual desire and performance. However, in a tiny minority of people these medications have a rather interesting side effect. Some of them have an orgasm whenever they yawn. I guess that this would be one group of people who would be happy to be cornered at a party by a really boring person …

Here's another weird link. Stretching of arms spontaneously when yawning can lead to a bizarre phenomenon in some stroke victims. A stroke occurs when part of the blood supply in the brain is blocked, causing localised damage to the brain and possible paralysis on one side of the body. Partially paralysed people cannot voluntarily move the arm on the affected side, no matter how hard they try. However, in 1923, the British neurologist Sir Francis Walshe noticed that when some stroke victims yawned, they could move their otherwise paralysed arm.

And here's another strange link. In heart attack victims and in people recovering in intensive care units, a spontaneous yawn is a sign that the person is hopefully on the way to recovery.

A pilot of an RAAF search-and-rescue maritime patrol plane, a P3 Orion, told me of his extensive experience over the Southern Ocean in severe and very turbulent weather, flying very low, looking for missing sailors. When he began to get a little airsick, he started yawning uncontrollably — very deeply, about once every minute. The yawns made it difficult to talk on the radio. If he had to fly in bad weather for a long time, he would get so tired that he would nod off and have a microsleep for five seconds at a time. This was not the best time to have a snooze — flying at 500 kph some 30 m over the ocean while doing hard turns and pulling lot of G-forces.

It seems hard to believe that you could nod off in this kind of extreme flying. I wonder if these guys are so used to living on the edge, that they no longer get the adrenalin rush that would have terrified someone like me into being wide awake.

Why Do We Yawn?

People who suffer cerebral anaemia or particular problems with the medulla oblongata will yawn almost continuously. But, in general, we don't really know why we yawn.

People are most likely to yawn in the first hour after they have woken up or in the last hour before they go to sleep — or when they're tired, bored, or doing long boring repetitive tasks.

Yawn and
Medical Emergency

I was on a five-hour flight from Sydney to Perth when a call came over the plane's public address system asking for a medical doctor. While I was trained as a medical doctor, my most recent clinical practice had been many years earlier when I was a doctor at the Kids' Hospital (now known as the Children's Hospital) in Sydney.

My heart sank when I saw my patient – a stout, middle-aged man with all the signs and symptoms of a heart attack. There was a limited amount that I could do with the medical supplies on the plane. There was no defibrillator to kick-start his heart back to a regular rhythm if it became necessary. The aisle on the jam-packed plane was so narrow and he was so stout that it was next to impossible to perform CPR (Cardiopulmonary Resuscitation) if I had to. I sat with him and kept feeding him oxygen.

And then he yawned. My heart leapt with joy. My medical teachers had told me that a yawn was often the sign that a patient was recovering from a medical crisis. As I sat beside him for the last hour of the flight, his hands became less clammy, his colour became more normal and his chest pain and tightness gradually decreased. I was so happy when I handed him over to the ambulance officers at Perth Airport. Later he phoned me to tell me that he was much better.

Yes, yawning is correctly linked to tiredness and to boredom, although super-awake and super-interested people will sometimes yawn. Look carefully at athletes getting ready for a sports event. Having trained for ages for this one short event, they will be fully alert and well rested. And yet they will yawn before their race, boxing bout or parachute dive. Musicians about to perform will often yawn — perhaps because of nervousness. A Siamese Fighting Fish about to rumble with another Siamese Fighting Fish will yawn 300 times more often than usual and dogs preparing to attack will also yawn more frequently — perhaps in anticipation. Charles Darwin noted in 1873 that if a baboon is involved in a threat or a matter of passion it was more likely to yawn. These are all seemingly odd times for a yawn to erupt.

One theory suggests that yawning is associated with a transition from awake to asleep (and vice versa), from alert to bored (and vice versa) and into sexual arousal (and vice versa). Yawning is linked to changing from any one activity to any other activity.

Contagious Yawning

However, we still don't understand why yawning can be contagious.

For a while, the yawn scientists thought that a circuit in the brain called the Mirror Neuron System might be involved. The brain cells in this circuit start firing when you do a particular activity and sense that somebody else is performing the same activity. However (and here's the confusing part), they don't fire when you see somebody else yawn and copy them.

We do know that empathetic types (e.g. people who, say, express concern when you stub your toe) are more likely to 'catch' a yawn while 'cold fish' types are pretty much immune to yawning in synchrony. We also know that a yawn by a human being can be 'caught' by their pet dog or cat. Chimpanzees watching a video of other chimps yawning will also start to yawn.

And finally, we know that thinking, hearing or reading about yawning has a 50% chance of setting off a bout of yawning. So, perhaps if this story made you yawn, you can identify yourself as a kind and caring soul, and there's nothing boring about that.

Other Theories of Yawning

One theory of why we yawn relates to the veins that surround the muscles that open the jaws – the pterygoid venous plexus. Supposedly, blood low in oxygen accumulates there during boring activity (when you don't use your jaw muscles much). Because the brain apparently doesn't like having stagnant deoxygenated blood so close, it kicks a yawn into action. The yawn both moves the venous blood along toward the heart and brings a big lungful of new air on board.

Another theory suggests that yawning helps regulate your body temperature. In one study volunteer yawners held temperature packs against their foreheads while watching others yawn. Warm (46°C) packs and room temperature packs had no change on the contagious yawning rate of the volunteers (about 41% of the time). However, cold packs on the forehead dropped this rate to just 9%.

References

Gallup, Andrew C. and Gallup, Gordon G. Jr., 'Yawning as a brain cooling mechanism: nasal breathing and forehead cooling diminish the incidence of contagious yawning', *Evolutionary Psychology*, 2007, Vol 5(1), pp 92–101.

Hooper, Rowan, 'You may want to yawn before reading this', *New Scientist*, 30 June 2007, p 14.

Platek, S.M., et al., 'Contagious yawning: the role of self-awareness and mental state attribution', *Cognitive Brain Research*, 2003, Vol 17, pp 223–227.

Provine, R.R., et al., 'Yawning: no effect of 3-5% CO_2, 100% O_2, and exercise', *Behavioural and Neurological Biology*, November 1987, pp 362–393.

Provine, Robert R., 'Yawning: the yawn is primal, unstoppable and contagious, revealing the evolutionary and neural basis of empathy and unconscious behavior', *American Scientist*, November 2005, pp 532–540.

Schurmann, Martin, et al., 'Yearning to yawn: the neural basis of contagious yawning', *NeuroImage*, 15 February 2005, Vol 24, pp 1260–1264.

Halley-less Comet

Early in 2007, people in the Southern Hemisphere were lucky enough to get a visit from Comet McNaught, discovered in 2006 by the astronomer Robert McNaught. And following the rules, he was allowed to name it after himself — one of the advantages of being an astronomer.

I was tremendously excited to see this comet with my own naked eyes. It was spectacular, multicoloured and huge, about five times the size of the Moon — if you were lucky enough to see it out of town, away from big city lights and pollution.

And in 1994, I was also very excited to be on board a NASA jet carrying an infrared telescope down towards the Antarctic. Through the telescope I saw the huge holes punched into the face of the planet Jupiter by the fragments of Comet Shoemaker-Levy 9 discovered by Carolyn and Eugene Shoemaker and David Levy.

On the other hand, Halley's Comet which appeared in our skies in 1985–86 was different. First, it was not so impressive and second, it was not named after the person who discovered it.

History of Comets

Comets, which have been observed for thousands of years, appear briefly in the sky, often with a long luminous tail. Indeed, the word 'comet' comes from the Greek word *kometes*, which means 'hairy one', referring to the long tail. In ancient times, their sudden appearance was thought to indicate the arrival of calamitous events, such as plague, flood or the death of some big wig.

In the 4th century BC, Aristotle thought that the heavens were pure and harmonious. Therefore, he believed that comets could not happen in the perfect heavens, but rather in our imperfect upper atmosphere. He was wrong, of course. Five centuries later, Seneca, a Roman philosopher, thought that comets belonged in the heavens along with the planets. However, his ideas were ignored in the face of Aristotle's authority and reputation. It took the mathematical genius of the Danish astronomer Tycho Brahe to prove that comets are heavenly bodies that orbit our Sun.

Today, we think of comets as being 'dirty snowballs', collections of loosely bound dust and various ices (e.g. water ice and methane ice). When, during an orbit, the body of a comet (measuring from 5 to 50 km in diameter) gets close enough to the heat of the Sun, the ices melt, blasting away a tail that can extend hundreds of thousands of kilometres.

Comet Shoemaker-Levy 9 (SL-9)

The Shoemakers and David Levy discovered comet SL-9 on the evening of 24 March 1993, using a wide-angle telescope — a 45-cm Schmidt at the famous Palomar Observatory in California. Through their fairly small telescope it looked like an unusually long and very dense object. However, other astronomers using much larger telescopes saw about 22 very bright, individual spots 'strung out like pearls on a string' over some 2 million km. These fragments ranged in size from a few hundred metres to a few kilometres across, implying that the original comet, before it broke up, was perhaps 5 km across.

When the astronomers calculated the flight path of the comet, they realised that it would hit Jupiter. Astronomers with a sense of humour didn't say that it would hit Jupiter, saying instead that it would pass within 45 000 km of its centre — a distance smaller than the planet's radius. (OK, so the joke is lost when you have to explain it!)

The stuff that makes a comet ... a comet

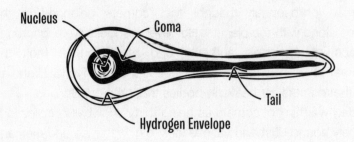

Nucleus

Coma

Tail

Hydrogen Envelope

A cross-section view of a comet

The orbit of Halley's Comet

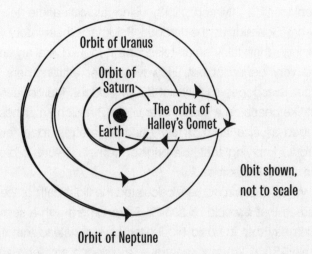

Orbit of Uranus

Orbit of Saturn

Earth

The orbit of Halley's Comet

Obit shown, not to scale

Orbit of Neptune

As the impact date grew closer, all available telescopes on Earth and in space swung towards Jupiter. The first fragment hit on 16 July 1994. It slammed into Jupiter at a speed of about 60 km/sec (200 000 km/hr), creating a spot on the surface some 6000 km wide. It also created a fireball with a temperature of approximately 24 000°C, the plume from this fireball rising about 3000 km above the surface.

Gravity Breaks up Comet

It's easy to understand how gravity can break up a comet. Think of the space shuttle and the Moon. The space shuttle flies around the Earth only a few hundred kilometres above the surface. It takes about 90 minutes for a complete loop around the Earth, travelling at about 28 000 kph. The Moon, much further away, takes about 29 days to do a complete loop, while travelling more slowly at 3500 kph.

Let's pretend that somehow, magically, the Moon vanished. Let's also pretend that at the same time, a huge comet suddenly appeared. This 'pretend comet' is so huge that it stretches from the orbit of the space shuttle to the orbit of the Moon. The side of this 'comet' closest to the Earth would try to do an orbit in 90 minutes. The far side of the 'comet' would try to do an orbit in 29 days. Both sides of the 'comet' would try to travel at different speeds and rip themselves off from the body of the 'comet'. Because it is so close to Earth, the 'comet' would immediately shred itself to pieces.

This is how the gravity of Jupiter broke up Comet Shoemaker-Levy 9, when it got too close to Jupiter. This could also explain where the Rings of Saturn come from. The gravity of Saturn broke up a bunch of large comets when they came too close and trapped the broken chunks in orbit.

Over the next six days, the remaining 21 fragments hit in 21 separate impacts. The biggest, Fragment G, created a scar on the surface of Jupiter approximately 12 000 km wide (roughly the size of Earth). This single impact released an energy equivalent to about 6 000 000 megatons of TNT — about 750 times the energy in all the nuclear weapons currently on Earth.

Luckily, Halley's Comet doesn't look like it will hit the Earth in the foreseeable future.

Shoemaker-Levy 9 was Weird

The comet Shoemaker-Levy 9 was a very weird comet because it was the very first comet to be discovered that did not orbit the Sun. When it was discovered, it was orbiting Jupiter instead.

Astronomers believe that, like all other comets, it originally orbited the Sun. But somewhere between the mid-1960s and the early 1970s, it came so close to Jupiter that it got gravitationally captured by this planet.

Jupiter, named after the Roman King of the Gods, is the largest planet in the solar system. In fact, Jupiter is so massive that, by itself, it accounts for more than 70% of the mass of all the planets in our solar system. At its centre, Jupiter reaches 20 000°C (hotter than the surface of the Sun), but its surface temperature is −114°C.

Jupiter is about 150 000 km in diameter, about one-tenth the size of the Sun or about 10 times bigger than our little Earth. Basically, it is a giant ball of gas with a solid core a bit bigger than our entire planet.

The core is solid because of the pressure of the overlying layers. The solid core is probably rock and ice, although scientists have also suggested exotic materials like solid metallic hydrogen and even diamond. However, we can't see the core because of the thick atmosphere.

On Earth, the atmosphere is about 10 km thick. But on Jupiter, the atmosphere is about 60 000 km thick – and very similar to the Sun (about 90% hydrogen and 10% helium) with traces of ammonia, methane and unstable compounds of arsenic and phosphorus. We call the top of the thick atmosphere the 'cloud tops'. One odd thing that can be seen in the cloud tops is the Great Red Spot, first observed by the English astronomer Robert Hooke in 1664. The Great Red Spot is a giant cyclone spinning anticlockwise and big enough to swallow three Earths side by side – and it's been there for at least 330 years.

Jupiter is so heavy that, after the Sun, it's probably the main 'vacuum cleaner' in our solar system, sucking up any random bits of matter that come flying through.

Halley the Megabrain

Edmund (or Edmond) Halley, a real intellectual heavyweight, lived from 1656 to 1742. In 1678, after spending two years in the Southern Hemisphere, he published *Catalogus Stellarum Australium*, the very first star catalogue of the southern stars, with their positions accurately measured by telescope. He mapped 341 stars, observed Mercury making a transit across the Sun and noted that some stars seemed to have become less bright over the ages. This assured his reputation as an astronomer, even though he did not yet have a degree. In the same year, he was also admitted to the Royal Society, the oldest and most prestigious science society in the world. King Charles II personally intervened to get Oxford University to grant him a Master of Arts degree.

In 1684 he was trying to prove Kepler's Laws of Planetary Motion, as well as working out exactly what forces kept the planets in their orbits around the Sun. So he visited the genius Sir Isaac Newton. To his surprise, he found that Newton had already solved this problem without telling anyone. Halley convinced Newton to write *Principia Mathematica*, one of the fundamental texts in physics, and even published it at his own expense in 1687.

In 1686, Halley published the first meteorological chart, showing the distributions of the winds over the oceans. He correctly identified the heat of the Sun as the driving engine for the winds. He also worked out the link between local air pressure and your height above sea level.

In 1690, he designed and built a diving bell and was able to stay underwater at 20 m for 90 minutes. He extended this time to three hours with a later design.

In 1693, he laid down the foundations of life insurance. He published one of the first tables relating death rates and age, using the meticulous mortality data from the Polish-German city of Breslau.

In 1698, he commanded the first sea voyage undertaken for purely scientific purposes, on the war sloop *Paramour Pink*. In his two-year journey, he travelled between 52°S and 52°N. In 1701, he published the first magnetic charts of the Pacific and Atlantic Oceans, showing the difference between True North and Magnetic North. These were absolutely essential for accurate navigation, and the charts were used for centuries after his death. Even today, these lines are called Halleyan or isogonic lines.

In 1703, Halley was appointed Savilian Professor of Geometry at Oxford University. Halley was a genuine megabrain.

And getting back to Halley's Comet, in 1705, he published his book, *A Synopsis of the Astronomy of Comets*.

Halley's Comet

Halley's book on comets examined the orbits of 24 comets between 1337 and 1698. He noted that three of them — the comets of 1531, 1607 and 1682 — had orbits that were so similar that they had to be the same comet. He correctly predicted that this particular comet would return in 1758. Indeed, Halley's Comet was the first comet to have its return correctly predicted.

He was dead when it did return, but in his honour, the comet was named after him, even though he did not discover it. In fact, our very first sighting of this comet dates back to 240 BC, about two millennia before his time. Halley may not have discovered Halley's Comet — but, like this Celestial Body, he definitely left an impression.

Significant Visits of Halley's Comet

2476 BC	Possible Chinese observation.
240 BC	Definite Chinese observation.
837 AD	Closest approach, at 5.15 million km, filling 90° of the sky.
1066 AD	Seen in England, and considered a bad omen, with Harold II dying at the Battle of Hastings. Its appearance was shown on the Bayeux Tapestry and was also recorded by Native Americans in their petroglyphs.
1301 AD	A further observation. The artist, Giotto di Bondone, depicted a comet as the Star of Bethlehem in his 1305 painting of the birth of Christ.
1986 AD	The least favourable visitation in all of human recorded history, because of its great distance from Earth and high urban light pollution. Because of its orbit, it was almost invisible in the Northern Hemisphere.
2061 AD	The next visit ...

References

Del Re, Gerard, *The Whole Truth: A Compendium of Myths, Mistakes and Misconceptions*, Random House Reference, 2004 (ISBN: 0-375-72066-9).
Encyclopaedia Britannica, Ultimate Reference Suite DVD 2006 — 'comet', 'Halley Edmond' and 'Halley's Comet'.

Psychic Sleuth

I have always had an uncannily uncomfortable feeling about the use of psychics as detectives — the ones who claim to use 'supernatural' powers to right wrongs. Sure, there are lots of reports about psychics who help law enforcement agencies solve crimes successfully. And yes, there are all those TV shows that involve psychic sleuths. But when I finally roused myself to look at this topic thoroughly, I found that it was all froth and bubble.

Psychics on TV

Some of the TV shows where psychics solve crimes for the police include 'The Dead Zone', '1-800-Missing' and 'Medium'.

In 'Medium', the actress Patricia Arquette plays a character based on Alison Dubois who lives in Phoenix Arizona. Alison Dubois hit the big time when she was noticed by Kelsey Grammar (who plays Frasier Crane in the TV sitcom, 'Frasier'). Besides being an actor, Kelsey Grammar produces 'Medium', the TV show inspired by the books of Alison Dubois, which has since become fabulously successful. The 'Medium' web site claims that 'Alison donates her time for missing/murdered persons and criminal cases for agencies across the country. She is contacted by law enforcement (agencies) and families to help find missing and murdered people.' The Glendale Arizona Police Department and the Texas Rangers are quoted as organisations that she has helped.

There are two problems with this.

I phoned the psychic.
He asked, 'Why are you calling?'
I said, 'You tell me.'

He's definitely dead ... and has two arms ... right? He also liked to drink water ... am I right?

First, both the Glendale Arizona Police Department and the Texas Rangers deny that she has ever helped them.

Second, in 2005–2006, two serial killers in Phoenix Arizona killed 11 people and wounded 16. If Alison Dubois is so good, why didn't she solve these crimes that happened in her own back yard?

The case unravels further when you consider that both the FBI and the National Center for Missing and Exploited Children state that psychic detectives have never solved a single missing person case in the USA.

How They Cheat

How do psychics get away with it? By using a few simple tricks.

First, they usually make vague predictions about the location of a body such as 'shallow grave in a forest'. Now ask yourself, would a killer ever dig a deep grave for their victim in an open field near a highway?

Second, they make really vague claims that they later reinterpret as specific, *after* the capture of the criminal. In one case the police arrested a murderer who drove a cement truck. The psychic then claimed that the cement truck was directly linked to her vision of a circle and that she had been on the case all along. In another case, the phrase 'don't walk alone' was interpreted to mean that the victim was 'confined to wheelchair'.

Third, they will dishonestly claim that wrong predictions are right. For example, there are two common and very different injuries to the brain. One is a brain haemorrhage, which is like a burst pipe. The other is a blood clot, which is like a blocked pipe. A burst pipe is very different from a blocked pipe. But some psychics will baldly claim that these two diseases are identical, even though they are very different.

The fourth trick is just a simple lie. Psychics will make a 'prediction' after the crime is solved but then claim that they actually made it beforehand. This is called 'retrofitting'.

Fifth, psychics always use the 'shotgun' technique where they spray out a huge amount of information, some of which must hit the mark. After all, you have a 50% chance of being correct if you say that someone is 'alive' or 'dead', or guessing if a baby will be a boy or a girl!

Another technique involves reading newspaper files and studying maps. For example, if highways 121 and 45 run near where the victim was last seen, a psychic has been known to state that the numbers 21 and 45 are significant. (They won't say '121' because skeptics would say that the psychic got this from a map. The number '21' is close enough to 'prove' that the psychic is indeed psychic.)

Bad News

In the USA, the Klaas family was shocked when their daughter, Polly, was abducted. Psychic detectives became involved, their interference proving counterproductive to finding the child. Marc Klaas now sees psychic detectives as 'the vanguard of a second wave of predators that also includes tabloid journalists, cheesy defense lawyers and photo-op politicians'. And their anguish was not lessened when one such psychic lied in claiming that she had helped the family solve the child's disappearance.

Psychics are as surprised as the rest of us when their plumbing overflows, their car engine explodes, relatives drop in suddenly or their plane is delayed. And while casinos do ban the use of mathematical techniques in certain card games, they have never banned psychics.

So why do psychics do it? Publicity has a value — which is why public relations firms exist. Psychics love this publicity, which leads to an increased demand for their palm readings, astrological predictions and other charming services.

A good example of this behaviour is that of the famous American psychic detective, Carla Baron. She claimed to have been engaged by the families in both the O.J. Simpson and JonBenet Ramsey cases — but this was later denied by the families.

And why do the cops occasionally use psychic detectives? Sometimes they want to cover up the real sources of their information. Sometimes they just want to unnerve superstitious suspects. Or perhaps the police are just getting into showbiz for themselves ...

Words, words, words ...

The word 'psychic' was first used by the physicist William Crookes to describe Daniel Dunglas Home (1833–1886), a famous magician and medium who claimed that he could talk with the dead.

By the way, 'psychic' as an adjective refers to the forces and agencies of a paranormal nature, while as a noun, it refers to a medium or a person who has psychic powers.

And as another aside, those who use this purely as a circus act and claim no paranormal involvement are called 'mentalists'.

References

Klaas in 2003, KlaasKids Foundation, http://www.klasskids.org/index.html.

Nickell, Joe, 'Investigative files: the case of the "psychic detectives"', *Skeptical Inquirer*, July 2005.

Nickell, Joe, 'Psychic sleuth without a clue', *Skeptical Inquirer*, May 2004.

Riccardi, Nicholas, 'Police arrest two in Arizona killings', *Los Angeles Times*, 5 August 2006.

Sleepy Time High

All parents of teenagers have seen the changes that happen to their kids with puberty, e.g. the growth spurts and the mood changes. The teenagers also acquire the virtually infinite ability to sleep in on weekends.

This altered sleep pattern has nothing to do with the teenagers being bone lazy or antisocial. Instead, it is caused by their changing biology lurking deep inside their brains. And for this reason I think that for teenagers, the school day should start later …

School Hours

In the USA, schools can start as early as 7.00 am. Much of continental Europe has 8.00 am start times, while the UK and Australia have 9.00 am starts.

However, two Victorian schools that have later starting times are both happy with the results. The Berengarra School in Glen Waverley now runs from 9.45 am to 3.00 pm while the Victorian P–12 College of Koorie Education in Mildura opens from 10.00 am to 3.40 pm.

Sleepy Time

How much you sleep depends on your age, as well as your maturation and development. Age is easy to measure — just use a calendar. The beginning of puberty (i.e. when the reproductive system matures) is fairly easy to mark. For example, hair starts to grow in the armpits and groin. The end of puberty is also easy to define. Puberty lasts until about 17.5 years of age in boys, and 16 years in girls, as measured by the end of bone growth.

However, adolescence (which occurs from 13 to 20+ years of age) continues for longer and is a lot harder to measure. In fact, recent research suggests that the easiest way to identify the end of adolescence is when the young person suddenly begins to need less sleep. This happens, on average, at 19.5 years of age in women and 20.9 years in men (according to a European survey of 25 000 people).

Newborn babies will sleep, if you add up all the naps, for 16–18 hours per day. By the age of five, this reduces to about 11 hours and continues to drop with age until puberty and adolescence start. Sleeping time then increases again until the end of adolescence.

Inside a teenage brain

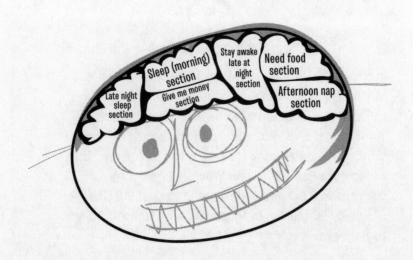

Adolescence, which covers the entire awkward time between childhood and adulthood, means independence from the family and the increasing importance of peer group activities outside the home. And during adolescence, the natural circadian rhythm is mightily interfered with.

First, there is a distressing delay in the onset of falling asleep, probably due to the later release of melatonin. I remember lying in bed as a teenager, listening to my parents' chiming clock mark away the night, before I would eventually drop off to sleep. Sometimes I would hear it run through 12 sets of quarter hours — three hours — before I finally fell asleep. When teenagers say that they are not tired at 11.00 pm, they are usually being truthful.

With the second biological change adolescents need more sleep — 9–10 hours every night. Indeed, one marker of the end of adolescence is the switch to adult sleep hours which are both shorter and earlier.

All of this means that the teenage years are very messy in terms of sleep. There are two problems — they can't get to sleep early and they need more sleep.

More Sleep for Adolescents

For adolescents an early bedtime is, in most cases, simply biologically impossible. But they still have to get up early for school. So they become sleep deprived.

Adolescents need 9–10 hours of sleep but often have to start school early, even earlier if they do sports. I remember getting up at 4.30 am to drive my son and daughter to rowing training. (Rowing is much easier, and a lot more fun, if the water is smooth and unbroken, as it usually is before the Sun rises, and before work and pleasure boats start running.)

You can easily see the effects of this sleep deprivation in an adolescent when the holidays finish. When the school term starts, they will sleep for two hours fewer on weeknights and try desperately to make up for it on the weekends. The result is that the average high school student is sleepwalking through the school day in a semi-permanent state of sleep deprivation.

Various studies have shown that this sleep deprivation can have nasty side effects. It is linked to rebellious behaviour, depressive symptoms, cigarette smoking, obesity, anxiety disorders and poor school marks. Indeed, being tired while taking an IQ test can drop seven points off your score — and can drop performance in regular exams. Just 23 minutes less sleep on a regular basis can drop an A student to a C student.

Sleep deprivation kills adolescents as well. On one hand, about 20% of all road deaths are caused by micro-sleeps, in turn caused

School Start Later?

From one point of view it makes sense for teenagers to start school later. And it's kind of easy to achieve. Primary school starting times remain the same – only secondary school starting times would be adjusted. And yes, it means that for the five or six years of high school, the students will be performing at their best.

But what about the social disruption to everybody else?

Parents with both primary and secondary school students will have to make two trips, unless the kids take the bus. Bus companies will have to spread their operating hours over a bigger window, both in the mornings and in the afternoons. (This could be an advantage, as fewer buses would probably be needed.) What about parents who drop the kids off at school on their way to work? Will they be able to negotiate with their employer to start work later?

If we do nothing, the students will spend most of their high school years being sleepy and getting worse marks than they should.

by tiredness. And about half of all such fatigue-related road deaths happen to those aged 16–25.

Dr Mary Carskadon from the Brown Medical School in Providence, Rhode Island, carried out a study on so-called REM-sleep (REM stands for Rapid Eye Movement). In this strange phenomenon your eyeballs flicker from side to side under your eyelids as though you are watching an invisible tennis match. During REM-sleep you do most of your dreaming, usually about 70–100 minutes after falling asleep. But half of the high school students in the study were so tired that when given the opportunity to sleep at school in mid-morning, they dropped into REM-sleep within a few minutes, not an hour and a half later. In other words, about half of the students showed symptoms of narcolepsy! None of the students in this study had a paid part-time job.

Other studies have shown that students who earn money by working 20 hours or more per week are even more exhausted and sleep deprived.

Cure Sleep Deprivation

So what's the cure? The first thing to realise is that adolescents are inherently good sleepers because they have robust circadian and homeostatic processes.

Therefore, adolescents should avoid caffeinated drinks, which mess up their circadian rhythms. And they should not have a computer or TV in their bedroom, which also encourages them to stay awake longer.

However, the real cure may involve society adjusting to their altered sleep patterns. Dr Martin Ralph, a psychologist at the University of Toronto, simply recommends that universities and high school classes start at 11.00 am. The kids can stay up later, get their needed sleep and then be wide awake in class.

And the next time you want to chide your teenager with a well-worn cliché in the interests of scientific credibility you'd better not make it the 'Early to bed and early to rise' one …

Effects of Sleep Deprivation

The Take Home Message is simple. You need sleep!

One study looked at more than a million people for six years. Those who slept less than 4.5 hours per night had a very high mortality rate.

In another study, Professor Eve Van Cauter from the University of Chicago restricted the sleep of healthy young male volunteers to four hours per night for six nights. This was enough to push their bodies into a pre-diabetic state. Dr Karine Spiegel has carried out similar studies with similar results. She says, 'Chronic short sleep is the royal road to obesity and diabetes'.

Lack of sleep also affects the immune system. One study looked at people getting flu shots. The sleep-deprived people got fewer benefits from their vaccinations.

References

Abbott, Alison, 'An end to adolescence', *Nature*, 6 January 2005, p 27.

Brink, Susan, 'Rest: It's required', *Los Angeles Times*, 9 October 2006.

Edwards, Hannah, 'Push to give sleepy-headed teens later class start time', *The Sun-Herald*, 14 January 2007, p 3.

Hansen, Martha, et al., 'The impact of school daily schedule on adolescent sleep', *Pediatrics*, 6 June 2005, pp 1555–1561.

Jenni, Oskar G., et al., 'Sleep regulation in adolescents', *Sleep*, November 2005, pp 1446–1453.

Millman, Richard P., et al., 'Excessive sleepiness in adolescents and young adults: causes, consequences, and treatment strategies', *Pediatrics*, 6 June 2005, pp 1774–1786.

Roenneberg, Till, et al., 'A marker for the end of adolescence', *Current Biology*, 2004, Vol 14, No 24, pp 1038–1039.

Rosen, Gerald, 'The value of a model', *Sleep*, November 2005, p 1345.

'School change for sleepy teens', *Courier Mail*, 4 May 2007.

Vince, Gaia, 'It came from another time zone', *New Scientist*, 2 September 2006, pp 40–43.

The CIA
Made AIDS?

A 2005 study examined the opinions and attitudes of African–Americans to HIV/AIDS. More than half of the people surveyed believed that HIV is an 'artificial virus' made for an evil purpose. More than half of the African–Americans surveyed also believed that there is a cure for AIDS — but that it is being deliberately withheld.

More specifically, about 15% of them believed that the CIA created HIV so that the American government could kill off the African–American population. Following the same line of thinking, Dr Wangari Maathai, the 2004 Nobel Peace Prize winner, said: 'Do not be naïve. AIDS is not a curse from God to Africans or the black people. It is a tool to control them designed by some evil-minded scientists.' (She has since 'clarified' her position.)

Difference between HIV and AIDS

First, let me be very clear about the difference between HIV and AIDS.

HIV (Human Immunodeficiency Virus) is a virus that attacks our immune system. There is actually a whole family of different types, varieties and strains of HIV. HIV-1 (the most common version in human beings) mutates very rapidly because it lacks 'error correction'. Whenever any DNA or RNA replicates and makes a few copies of itself, it always makes mistakes. In most living creatures, there are 'error correction' or 'ploofreading' (joke!) functions that find and repair the errors. Because HIV-1 does not have 'error correction', it evolves one million times faster than human beings do. This makes finding vaccines or drugs to fight HIV-1 an uphill task. It also means that HIV-1 can mutate quickly into a form that the human immune system does not recognise as foreign, thus escaping attack by our immune system.

On average, a mature HIV particle has a diameter of about 120 nm (billionths of a metre). You can visualise the individual virus particles with an electron microscope.

We humans might think that we are the peak of the evolutionary tree, but as far as bacteria, viruses, fungi and parasites are concerned we are food, or a convenient home. Most of the time, our immune system is active enough to fight off these potential invaders. However, when our immune system is weakened, these various germs invade and take over our body.

AIDS (Acquired Immuno-Deficiency Syndrome) is the disease caused by the virus known as HIV. It causes the injected person's immune system to become depressed and therefore unable to fight off infections. In Stage I, which is not classified as AIDS, there are actually no symptoms but the virus is somewhere inside their body. In Stage II there are minor lesions on the skin and mucus membranes and recurrent upper respiratory tract infections. In Stage III the symptoms include tuberculosis in the lungs, nasty bacterial infections and unexplained chronic diarrhoea for longer than a month. In Stage IV infectious agents that are usually kept at bay by a healthy immune system run rampant through the victim's body. The diseases that these infections cause include candidiasis of the oesophagus and upper airways, Kaposi's Sarcoma and toxoplasmosis of the brain.

HIV/AIDS Appears

On 5 June 1981, the Center for Disease Control in the USA released its regular *Morbidity and Mortality Weekly Report* which discussed the case of five young gay men suffering from a strange set of symptoms. Their disease turned out to be AIDS. Since then 25 million people have died from AIDS and 65 million people have been infected with HIV, the virus that causes AIDS.

Even many years later, there are still many strange notions about AIDS and HIV.

HIV Timeline

1959
Preserved blood sample carrying HIV, taken from an adult Bantu male living in Kinshasa in the Democratic Republic of Congo.

July 1981
Five young gay men with a strange cluster of symptoms come to the attention of the CDC (Centre for Disease Control). Their illness is given the name GRID (Gay Related Immune Deficiency).

July 1982
The illness is renamed AIDS (Acquired Immune Deficiency Syndrome), because people other than gays were now becoming infected. These people included female sexual partners of men with the disease, haemophiliacs and other blood transfusion recipients, and children of infected mothers.

1987
About 36 000 people (men, women and children) now diagnosed with AIDS with 25 000 dead from the illness.

2006
About 25 million people killed by AIDS so far and about 38 million people infected and living with HIV. Each year five million people become infected and three million die. About 70% of the infections happen in sub-Saharan Africa.

Where HIV Came From

Molecular biologists have actually traced the natural origin of HIV. After decades of painstaking work analysing the genetic code of various relevant viruses that infect monkeys, chimpanzees and human beings, the true picture is finally emerging. It relates to what is called 'bushmeat', the eating of wild animals. It also relates to viruses that have been infecting monkeys and chimps for thousands of years but, strangely, don't really harm them. Monkeys and chimps have been the natural hosts for various viruses for centuries — and simply don't become ill from the infections.

The true origin of HIV began with two separate viruses that infected two separate species of monkeys. The Red-capped Mangabey monkey (*Cercocebus torquatus*) was infected by SIV_{rcm}, while the Greater Spot-nosed Guenon monkey (*Cercopithecus nictitans*) carried SIV_{gsn}. (SIV stands for Simian Immunodeficiency Virus — the word 'simian' meaning 'monkey' or 'ape'.) These monkeys were then attacked and eaten by chimpanzees of the subgroup *Pan troglodytes troglodytes* who, because of their superior size (up to 1.8 m), strength and clever organisation into formidable teams, easily hunted down and ate the monkeys, while they were alive — a ghastly business indeed. During the attack there was lots of monkey blood splashed around, and occasional cuts to the chimpanzees, so the chimps became infected with both monkey viruses. Inside the chimps, the two monkey viruses mutated into one new virus — SIV_{cpz}. This new virus did not make the chimps unwell.

Somewhere between 1910 and 1950, a human being hunted down and killed a chimp carrying SIV_{cpz} for 'bushmeat'. The chimp virus, SIV_{cpz}, in the chimp blood which entered his body through cuts, infected him and then mutated into HIV. HIV is a so-called 'lentivirus', i.e. it is a slow growing virus with a typical incubation period of 10 years before the full-blown symptoms arise.

The genetic code in all of these viruses leaves a clear trail of the evolution that began with two separate monkey viruses, had an intermediate stage in chimpanzees and ended up with HIV in human beings.

Conspiracy Theory #246: The CIA made AIDS

A 2005 study looked at the opinions and attitudes of African-Americans to HIV/AIDS. Over half of the people surveyed believed that HIV is an 'artificial' virus, made for evil purposes.

How it happened

In humans, SIVcpz turned into HIV

Chimpanzees

SIVcpz

 SIVrcm

SIVgsn

Monkeys

In fact, this transmission of virus from chimp to human being has almost certainly happened not just once but many times.

Different Varieties of HIV

Two strains have been identified – HIV-1 and HIV-2. Both strains cause AIDS.

HIV-1 is much more common, more deadly and more easily transmitted than HIV-2. It has many versions – three groups (M, N and O) and 10 subtypes (A–K). It appears that HIV-1 infected human beings via a complicated pathway.

It began in two monkey species – one carrying SIV_{rcm}, the other carrying SIV_{gsn}. Transferred to chimpanzees, the two viruses combined to form a new virus – SIV_{cpz}. This virus then infected human beings (via cuts and scratches or the consumption of bushmeat), finally evolving into HIV-1.

Harder to catch, HIV-2 is less virulent than HIV-1. It almost certainly began with the SIV_{sm} virus in the Sooty Mangabey monkey (*Cerocebus atys*) or the SIV_{agm} virus in the African Green monkey (*Cercopithecus aethiops*). It infected human beings following a similar pathway to that taken by HIV-1.

Making Viruses

One of the reasons why the CIA (or Mossad, or the National Security Agency, or ASIO, or MI5, or the KGB) could not have made HIV is very straightforward. A quarter of a century ago we simply did not have the knowledge to make such a virus.

We have developed the knowledge and technology to make a complete working virus only in the past few years. We didn't develop PCR (which lets us easily make billions of copies of bits of DNA that we are interested in) until the mid-1980s. And we didn't develop the ability to 'read' the genetic code of viruses until the 1980s. We needed very high-level molecular biology skills to make a virus — and we have only recently developed these skills.

As a result, the polio virus wasn't made until 2002. The polio virus is quite small (with only 7741 base pairs). Even so, it took two years of work.

The influenza virus of 1918 that killed 50 million people worldwide took even longer — it was reconstructed in 2006. It is a much larger virus with 2.8 billion base pairs, similar to the human DNA, which has 3 billion base pairs.

However, we have a preserved blood sample, taken in 1959, that shows that HIV was infecting people even then. We definitely did not have the technology to make HIV in 1959.

Conspiracy Theories

There are many conspiracy theories about AIDS/HIV. The main ones are:
- The virus was made by Big Government to kill off undesirable citizens.
- It was transmitted by a tainted oral polio vaccine of the 1950s.
- It is God's punishment for the 'sin' of homosexuality.

Why Believe a Crazy Myth?

So why do people, especially African–Americans, believe such a preposterous myth?

There are a few reasons.

The Soviet newspaper *Literary Gazette* published an article in 1985 which claimed that HIV was created in a US Army research laboratory at Fort Detrick in Maryland. However, its only source for this claim was a leftist Indian newspaper, *The Patriot*.

A second and very good reason is the well-documented Tuskegee Syphilis Study of Alabama. From 1932 to 1972, about 400 poor and mostly illiterate African–Americans with syphilis were studied in order to understand the natural course of the disease. However, when the simple cure of penicillin became available in 1947, it was deliberately withheld from this group. The syphilis made their lives short and miserable, leaving most of them with horrendous, subsequently fatal disabilities. It also infected and killed spouses. The 'study' continued until 1972, when it was leaked to the press. It was not stopped for ethical reasons.

A third reason is the long history of maltreatment that African–Americans have suffered at the hands of the ruling Whites. They have endured two centuries of slavery followed by more than a century of oppression. Why should they believe that things have changed? A degree of paranoia is perfectly reasonable considering the circumstances.

But there is still one mystery that has not been solved. How is it that the original virus infects but does not really harm monkeys and chimpanzees? Perhaps understanding this mystery will further reduce the number of deaths from AIDS …

A Lousy Way to Kill People

If someone wanted to use a virus to kill people, then HIV is probably one of the least effective to use.

First, this virus is quite difficult to transmit sexually. It will pass from human to human only 1–10 times per 10 000 sexual acts. This is much less efficient than other STDs (Sexually Transmitted Diseases).

Second, the virus takes a very long time to kill people. It takes about 10 years from the first infection for the victim to start suffering the symptoms of AIDS and another two years for them to die (if the disease is left untreated). People are usually perfectly well during this

10-year window, which means that they can unknowingly infect other people.

In general, the kinds of bio-terror weapon that the military seems to prefer and spend their money developing are quite different from HIV. They tend to have a short onset, are severe and highly contagious. They also have a 'release' that is controllable. Finally, the nasty perpetrators who released it in the first place have some kind of treatment or vaccine, in case the disease comes back to infect them. Typical bio-terror infections include Ebola virus, smallpox and anthrax.

Anyone wanting to use a virus to kill a group of people wouldn't have to invent a new virus – there are plenty of perfectly nasty viruses with much more suitable and toxic characteristics already available.

References

Bailes, E., 'Hybrid origin of SIV in chimpanzees', *Science*, 13 June 2003, p 1713.

Bogart, L.M., et al., 'Are HIV/AIDS conspiracy beliefs a barrier to HIV prevention among African Americans?', *Journal of Acquired Immune Deficiency Syndrome*, February 2005, pp 213–218.

Cello, J., et al., 'Chemical synthesis of poliovirus cDNA: generation of infectious virus in the absence of natural template', *Science*, 9 August 2002, pp 1016–1018.

Curtis, Tom, 'The origin of AIDS: A startling new theory attempts to answer the question "Was it an Act of God, or an Act of Man"', *Rolling Stone*, 19 March 1992, pp 54–59, 61, 106, 108.

Gao, F., et al., 'Origin of HIV-1 in the chimpanzee Pan troglodytes troglodytes', *Nature*, 4 February 1999, pp 436–441.

Gibbs, Walter, 'Nobel peace laureate seeks to explain remarks about AIDS', *The New York Times*, 10 December 2004.

Heeney, J.L., et al., 'Origins of HIV and the evolution of resistance to AIDS', *Science*, 28 July 2006, pp 462–466.

McNeil, Donald G. Jr., 'Researchers have new theory on origin of AIDS virus', *The New York Times*, 13 June 2003.

Taubenberger, J.K., 'Characterization of the 1918 influenza virus polymerase genes', *Nature*, 6 October 2005, pp 889–893.

Zhu, T., et al., 'An African HIV-1 sequence from 1959 and implications for the origin of the epidemic', *Nature*, 5 February 1998, pp 594–597.

The X-Chromosome eXplained

Back in the Olden Days, before research into cloning and stem cells, most people didn't know much about genetics or DNA. However, they had a vague impression that there was something called the 'X-chromosome', so named because it looked like the letter 'X'. Well, this is not the case, but as an aside, the X-chromosome did help give the world Communism.

Shape of DNA

Every cell in your body (except for the red blood cells) carries DNA. (Red blood cells are so dedicated to their job of carrying oxygen efficiently, that everything irrelevant has been stripped off including the DNA. Red blood cells are not made by other red blood cells — they are made by stem cells in the bone marrow in the long flat bones of your body.)

The human DNA is a very skinny and very long molecule. The DNA in each molecule is a few billionths of a metre wide, but if you stretched it all out it would be a few metres long. It looks like a ladder with two side rails and about three billion rungs joining the side rails to each other.

As you can see, DNA is shaped like a long, twisty, rope, laddery-type thing.

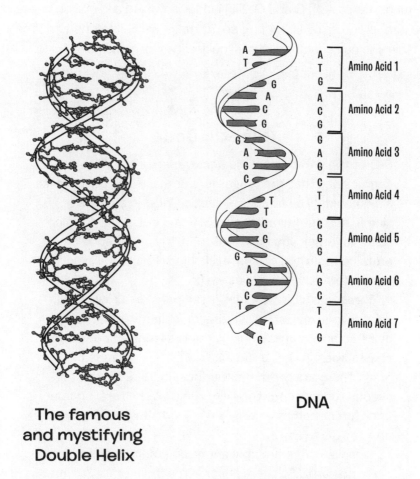

The famous and mystifying Double Helix

DNA

The human DNA is a very skinny, and very long, molecule. The DNA in each molecule is a few billonths of a metre wide, but if you stretch it out it's a few metres long. It looks like a ladder with two side rails and about three billion rungs joining the side rails to each other.

This ladder is twisted into a right-handed spiral as part of an efficient way of folding something a few metres long into a space smaller than one-millionth of a metre wide. The twisted ladder is about 2.3 nm (nanometres or billionths of a metre) wide. The scientific name for this molecular structure is a 'double helix'.

The 'rungs' are 0.34 nm apart. There are four different types of rungs, called A, T, C and G. The ladder is twisted and one complete 'turn' of the spiral is 3.4 nm, so 10 rungs will fit into one turn. The scientific name for the rungs is 'nucleotides' or 'base pairs'.

Genetic Code

One of the great scientific discoveries of the 20th century was that the rungs were actually a 'code' to make amino acids. The famous 'Genetic Code' is amazingly simple and yet incredibly deep. If you put enough amino acids together, you have a protein. If you put enough proteins together, you have a living creature (OK, you need a few other things as well).

Three rungs (nucleotides/base pairs) in a row have enough information to tell the 'machinery' in the cell to make an amino acid. Look at the first rung. There are four possibilities: A, T, C or G (the four different types of rung). There are the same four possibilities for the second rung, and for the third rung. So the total number of different combinations is 4 x 4 x 4 = 64 (running from AAA to TCG to GGG).

However there are only about 20 common amino acids in life on Earth. So there is some redundancy, i.e. several combinations of A, T, C and G will give the same amino acid. For example, the combinations CGC, CGA, CGG, AGA and AGG will all tell the 'machinery' in the cell to make the amino acid arginine. However, at the other extreme, two of the amino acids have only a single combination each, e.g. methionine and tryptophan.

Job of DNA

Our DNA is, among other things, an architect's blueprint that will make and then maintain a human being.

Most of the time, the DNA exists as a myriad of long slender filaments, floating all tangled up in the nucleus of the cell but not at all neatly condensed. This gives them a huge surface area, which makes it easy for the 'machinery' in the cell to 'read' the DNA to make proteins. These proteins could be insulin from your pancreas, enzymes from the cells in your gut to dissolve your food or muscle in your muscle cells to move your arms, legs and eyelids.

Chromosomes

Whenever a cell is about to split into two more cells, the DNA will condense for a brief time into little clumps. A skinny strand of DNA gets wound into a coil, and this coil gets wound again, and so on — a process called 'DNA supercoiling'. As part of this process, proteins are wrapped by the coiling DNA and also wrap around the DNA. These clumps that appear when a cell divides are the famous chromosomes. They have a central point and four arms, making them look a little like the letter 'X'.

By the way, the number of chromosomes varies with the species — just eight in the fruit fly, 46 in human beings and hares, 48 in gorillas and chimpanzees, 104 in goldfish and a massive 380 in butterflies.

Chromosomes were first seen in cells by the Swiss botanist Karl Wilhelm von Nageli in 1842. Chromosomes are really hard to see, but if you soak the cells with the right dyes you can then see these coloured bodies — hence the name 'chromo' meaning 'colour' and 'some' meaning 'body'.

The original technique to visualise chromosomes was to 'poison' the cell with a drug called colchicine, which locks the cell in at a certain stage of division. The scientists then stained the cell with dyes to make the chromosomes obvious, took a photograph (via a microscope) of the chromosomes, developed the photograph, cut out the chromosomes with scissors, arranged them in pairs and stuck them down with sticky tape. Nowadays, it's done with fancy, digital computer magic.

Shapes of Chromosomes

Chromosomes come in two main shapes.

The so-called Linear Shape is the 'classic' X-shape. But it doesn't really look like an 'X'. Instead of four separate arms all coming from a single point, there are usually two sets of parallel legs. Usually, there are two shorter arms (called 'p' from the French word *petit* meaning 'small') and two longer arms (called 'q', because 'q' is the next letter in the alphabet after 'p'). Like the Y-chromosome, this is another case of a name being chosen because it's the next letter in the alphabet.

The other chromosome shape is the circle. This is often found in smaller creatures, such as bacteria.

Mystery of Chromosomes

It took a long time to learn about our chromosomes. In fact, until 1955, we thought that human beings had 48 chromosomes — the real number is 46.

They were a huge mystery until very recently, the most mysterious of them all being the 45th chromosome. Part of the mystery surrounding this particular chromosome was its involvement in diseases carried by females. Although these diseases, which included haemophilia and red–green colour blindness, didn't affect females, they affected males — very strange.

For a long time, this mystery remained unsolved. In algebra, the symbol 'X' stands for the unknown quantity, as in the X-Factor — and this is how the X-chromosome was given its name. (If it was named after its shape, then all the chromosomes would be called 'X'.)

The chromosome

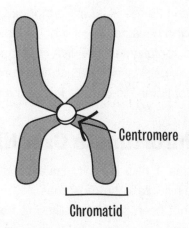

Centromere

Chromatid

Chromosomes come in two main shapes.

The so-called Linear Shape is the 'classic' X-shape. But it really doesn't look like an 'X'. Instead of four separate arms all coming from a single point, there are usually two sets of parallel legs. Generally there are two shorter arms (called 'p') and two longer arms (called 'q').

The other chromosome shape is the circle. This is often found in smaller creatures such as bacteria.

And the Y-chromosome? Well, it was pretty mysterious too. 'Y' is the next letter in the alphabet after 'X', which is how the Y-chromosome got its name. It's as simple as ABC.

Communism

So what's the link with Communism? Admittedly, it is a little tenuous but it is related to the X-chromosome.

It seems that Queen Victoria had a spontaneous mutation in her X-chromosome that could cause the bleeding disease, haemophilia. This led (via some very convoluted logic) to Communism.

Queen Victoria had nine children. Her third child, Alice, who carried this mutation in her X-chromosome, married Louis IV, Grand Duke of Hesse, in Germany and passed the haemophilia mutation to her daughter, Alix. Alix, who took the name Alexandra when she was baptised into the Russian Orthodox Church, married Nicholas Romanov, who became Czar Nicholas II of Russia. She passed the haemophilia mutation to her fifth child, and first son, Alexei. He suffered from debilitating haemophilia from an early age, not very favourable for someone destined to be the future Czar.

Unfortunate Czar Nicholas II

Czar Nicholas II had his fair share of worries — most of them unrelated to his wife's chromosomes.

He had not been properly trained to be the Czar of Russia. For example, when approached by a respectful delegation of peasants and workers who asked for some reasonable constitutional reforms to improve their wretched lives, he made long-lasting enemies by angrily rejecting them. If he had been diplomatically trained, he could so easily have responded carefully, making the peasants his lifelong allies.

On 28 June 1914, the Austrian Archduke Franz Ferdinand was assassinated in Sarajevo by a member of a gang called the Serbian Black Hand. In retaliation, Austria declared war on Serbia. Czar Nicholas II, an ally of Serbia, mobilised his army against Austria. Because Germany was an ally of Austria, he was now at war with two countries. Helping start World War I doesn't look so good on your CV.

Incredibly, things got even worse. His army lost many battles, four million Russians dying in the first year of battle alone. So he took over command of the army and lost even more battles. The difference now was that as Commander in Chief, he was personally responsible.

General unrest led to an outbreak of revolution in the capital, St Petersburg. But because Nicholas was away from the capital, he was unable to control it. And it didn't help that his wife Alexandra was a German — not a very popular nationality in Russia at the time of World War I.

All this led to enormous unrest and, ultimately, to the Bolshevik uprising.

Part of the Czar's inability to attend properly to his duties of office may have been his preoccupation with the terrible haemophilia of their only son, Alexei. Therefore, he failed in his regal duty to govern Russia properly. And that's how the X-chromosome played a mysterious part in giving the world Communism …

References

Angier, Natalie, 'For motherly X chromosome, gender is only the beginning', *The New York Times*, 1 May 2007.

Steven, Richard F., 'The History of haemophilia in the royal families of Europe', *British Journal of Haematology*, April 1999, pp 25–32.

She Smells Seasmells (Ozone) ...

As a child, I remember going to the seaside with my parents and being told to 'smell the ozone' as it was healthy and good for me. The belief that seaside ozone is good for you has persisted in our literature and, of course, in the advertising blurbs of companies selling ozone products that you just can't live without. In the 19th and early 20th centuries, there were ozone hospitals, ozone hotels and even ozone cages built at the seaside, specifically to get the healthy ocean/ozone air. Even today in the 21st century, the *Collins Dictionary* still defines 'ozone' as 'clean bracing air, as found at the seaside', while the *Macquarie Dictionary* calls it 'clear, invigorating, fresh air'. And the RAN clinic, a health resort in the Israeli city of Arad, offers the healthful effects of ozone as part of its range of natural therapies.

However, we've all been duped. Professor Andrew Johnston from the University of East Anglia in the UK says that 'we were misled, twice over. Firstly, because that distinctive (seaside) smell is not ozone, it is dimethyl sulphide. And secondly, because inhaling it is not necessarily good for you.' Although Professor Johnston was not the first person to note that the smell of the seaside is not ozone, he reminded us all of this. I certainly didn't know this until I read his work. He analysed the smells coming out of the mud of the Stiffkey saltmarsh on the Norfolk coast of Great Britain and showed, in molecular terms, how the bacteria made dimethyl sulphide (DMS).

Ozone

Our atmosphere consists of about 78% nitrogen and 21% oxygen. These oxygen molecules are made of two oxygen atoms stuck together. But there is also another form of molecular oxygen called ozone which has three oxygen atoms stuck together.

The odour of ozone was first noted around electrical machines in 1785. It is manufactured when an electrical discharge passes through regular molecules of oxygen with their two atoms and converts them into ozone molecules with three atoms of oxygen. It was discovered in 1839 by Christian Friedrich Schonbein, who made it by passing electricity through water. He called it after *ozein*, the Greek word for 'smell', because of the strange odour that you might sometimes smell in a lightning storm. The fact that it was made of three oxygen atoms was worked out in 1872.

Ozone has become famous over the past few decades via the 'hole in the ozone layer'. In the stratosphere about 10–50 km above the ground the ozone protects us from ultraviolet radiation. However, at ground level, high levels of ozone can irritate the eyes and mucous membranes.

We are not quite sure how the myth of 'healthy ozone' started. Perhaps people thought that because ozone had one extra oxygen atom, it was somehow better than the 'regular' oxygen molecules in air. (If two atoms of oxygen are good for you, then surely three atoms have to be 50% better!) Ozone has an odour similar to very dilute chlorine — and the seaside sometimes does have this odour as well. It seems that this similarity was enough to start the 'seaside ozone is good for you' movement, which reached its peak in the late 19th century.

DMS

However, the smell at the seaside is not even ozone. It's a very different chemical called dimethyl sulphide or DMS.

'I love the smell of ozone in the morning!'

The 'ozone' smell that we have all been led to believe is part of the smell of the beach is actually the smell of the chemical dimethyl sulphide (DMS). The sensation of the smell depends on the concentration. It is given off when you cook corn or cabbage. In high concentrations, it can smell a little like garlic. And in lower concentrations, it can smell like ozone.

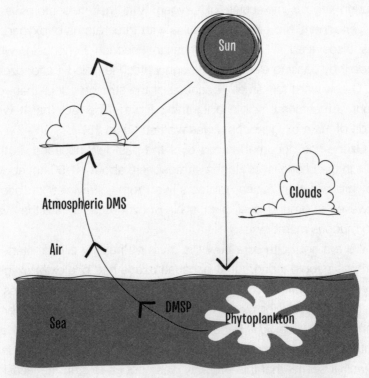

Phytoplankton release DMSP when they are attacked by bigger creatures.

Micro-organisms in the ocean then do their bit to convert DMSP to DMS, or that all too familiar 'ozone beach smell'.

DMS is a clear inflammable liquid that boils at 37°C but can exist as a gas at lower temperatures. It is given off when you cook corn and cabbage and in a different concentration is responsible for the characteristic smell of Brussels sprouts. In high concentrations, it smells a little like garlic. It is sometimes added to dangerous, odourless gases to make them obvious. And it sometimes appears as an unpleasant contaminant in beer. (The remedy here is to have good sanitation procedures and to maintain a rolling, uncovered boil when brewing the beer.)

This chemical has some industrial uses from enhancing corn flavours in food to refining petrol and controlling dusting in steel mills. It is also one of the main odours in black truffles (those delicious but expensive underground fungi), but it rapidly fades upon exposure to air.

DMS and the Ocean

In the ocean, many of the creatures make a chemical called DMSP which is then converted to DMS. These creatures then dump the DMS into the ocean around them. This might explain how some birds can find their next meal in the apparently featureless ocean. In fact, DMS is part of the natural 'smell-scape' of the Southern Ocean.

DMS can act as both a 'scream for help' and a 'homing signal'. For example, tiny phytoplankton release DMS when they are attacked by bigger creatures, such as zooplankton. Some species of crustaceans and seabirds (e.g. petrels and shearwaters) can sense very low levels of DMS and will swoop in, following the smell, to eat the bigger creatures while they are in the act of eating the plankton. The phytoplankton call for help and the petrel flies in and eats their attackers. What carnage!

Even tiny creatures like bacteria release DMS. In fact, a huge quantity of DMS is released from the ocean — about 30–50 million tonnes per year. The oceans make about 50% of all the biological sulphur that gets dumped into the atmosphere. In turn, about 90% of the sulphur that comes from the ocean is in the form of DMS.

Professor Johnston has been involved in understanding how the genes of various bacteria help them live, e.g. how they get nitrogen from the atmosphere. Recently, he and his team have been studying how bacteria make DMS. They found some genes that control the production of DMS — and, almost as an aside, casually mentioned that DMS could cause the seaside smell.

So, back beside the seaside, DMS in high concentrations is an irritant to the eyes and mucous membranes — but you're safe with the levels that you usually get at the beach. DMS is not the only component of the 'seaside smell'. Various other chemicals, including those coming from algae, ice cream and sunscreen, are involved. But ozone isn't one of them!

DMS Makes Clouds and Climate Change

DMS has achieved fame with scientists over the past few decades – not least because it can affect the climate on our planet. DMS rises from the ocean (where various creatures make it) and oxidises to make tiny floating particles of sulphate. In turn, these sulphate aerosols seem to be the major source of what are called CCN or Cloud Condensation Nuclei. These CCN trigger the production of clouds which can then reflect the sunlight and perhaps cool the Earth.

One paper suggested that if the numbers of CCN doubled, this could perhaps counteract the warming caused by doubling the amount of CO_2 in the atmosphere!

Perhaps this is just a fanciful suggestion. However, DMS does help control the climate – and the ocean does dump about 30–50 million tonnes of it per year into the atmosphere.

References

Charlson, R. J., et al., 'Oceanic phytoplankton, atmospheric sulphur, cloud albedo and climate', *Nature*, 22 April 1987, pp 655–661.

Nevitt, G. A., 'Dimethyl sulphide as a foraging cue for Antarctic Procellariiform seabirds', *Nature*, 24 August 1995, pp 680–682.

Todd, Jonathan D., et al., 'Structural and regulatory genes required to make the gas dimethyl sulfide in bacteria', *Science*, 2 February 2007, pp 666–668.

How to Make a Wee Difference!

When you were a kid, did your parents ever tell you to go to the toilet and wee 'just in case' before you went out? The 'in case' was to save needing to 'go' while you were in the car, out shopping, at the movies or at any other social function. This advice can lead to a habit of a lifetime — but it is not the advice that you would usually get from a doctor treating urinary incontinence.

Kidney to Bladder

The bladder is a hollow organ that is the reservoir for urine. It is located very low down in the abdomen, just above and behind the pubic bone.

The volume of urine that you pass depends on how much water you drink and how much you sweat. When you are asleep your urine production slows down to about 0.25 ml per minute, enabling you to get a good night's sleep. However, if you are anxious, worried or frightened, this mental state can increase your blood pressure, which means more blood through the kidneys. So you make more urine. In cold weather less blood goes to the skin. This also means more blood to the kidneys — and again, more urine. (This is why you wee more in winter when the blankets are so warm and the world outside is so cold.)

Urine is made in both kidneys at the rate of about 1 ml per minute. It then flows out via the two ureters, each about 25 cm long. The ureters have walls of muscle arranged in spiral and longitudinal layers. Every 10–60 seconds, waves of synchronised muscle contraction progress down the ureters, pushing urine into the bladder. (This motion is called peristalsis.)

Bladder Anatomy and Physiology

The bladder has a capacity of 500 ml (roughly a small carton of milk). And like the ureter it has various bands of muscle in its wall — spiral, longitudinal and circular. When the bladder empties the muscles contract in a specific sequence, starting at the top and working down to the bottom. This 'wrings' virtually all of the urine from the bladder.

Although you get your first urge to pass water with a volume of 150 ml of urine, you can usually ignore this. However, it's harder to ignore the marked sense of fullness that normally happens with a volume of about 400 ml. As the bladder fills, it gradually rises up into the abdomen, sometimes even reaching up as far as the umbilicus (or bellybutton).

The desire to urinate comes from sensors in the walls of the bladder that pick up the internal pressure. This pressure arises from having a full bladder or from having something pushing on a not-so-full bladder. Therefore, in pregnancy, when the baby's head pushes on the bladder, this leads to a desire to pass water even when there is only a small amount of urine inside it. Mental stress such as excitement or anger can tighten up the muscle wall leading to increased pressure and, once again, producing the desire to urinate.

Unstable Bladder

If you get into the habit of going to the toilet before your bladder is full, you can develop a problem known as 'detrusor instability' or 'unstable bladder'.

This is piss easy!

The bladder is a hollow organ that is a reservoir for urine. It's found very low down in your abdomen, just above and behind the pubic bone.

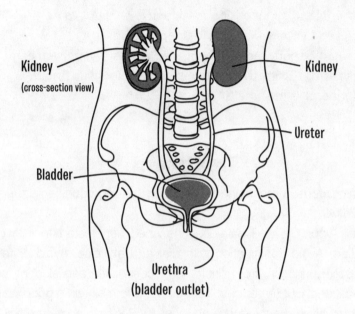

Kidney
(cross-section view)

Kidney

Ureter

Bladder

Urethra
(bladder outlet)

Urinary structures

By passing urine 'in case' you need to go later, you can accidentally 'reset' the sensors in the bladder. These sensors then fire off messages to the brain that you need to pass urine even though the bladder might be only partly full. This can lead to your feeling the urge to pass urine every 20 minutes or so. But frustratingly, when you go, you pass only a small volume. By going to the toilet too frequently you can perpetuate the cycle of triggering the desire to urinate. (Diabetics may also need to go to the toilet very frequently but usually pass large volumes of urine.)

Along with the sensation of urgency to pass urine, you can also develop urinary incontinence. This can cause a lot of embarrassment and inconvenience for sufferers.

Urinary Incontinence

For adults, detrusor instability or unstable bladder is one of the two common causes of urinary incontinence. The other is stress incontinence which is thought to be caused by weakened pelvic floor muscles.

Just to show how complicated things can be, you can also suffer from stress incontinence and an unstable bladder at the same time.

Treatment

The first steps in managing this very distressing condition are easy to follow.

Visit your GP and have a urine specimen tested to rule out infection. Avoid constipation with diet and exercise. Avoid diuresis (excessive urination) from drinks (e.g. coffee, tea and alcohol) and medications. But remember that you need to maintain a good water intake to 'flush' your kidneys adequately. Because some medicines have a diuretic effect, you will need to have your regular medications reviewed. Specialised bladder testing can confirm the diagnosis of an unstable bladder.

You Just Keep Me Hangin' On ...

People with 'unstable' bladders can find themselves wanting to go the toilet every 20–30 minutes. However, you should just sit back and literally hold on.

Just like five-year-old children physically clutching their groin when they feel that they are bursting to do a wee, you have to sit down, cross your legs and distract yourself mentally in an effort to ignore the first desire to pass urine. Wait for it to pass without rushing to the toilet. The idea is to gradually increase the time between each urination and retrain your bladder to registering that it is full only when it is actually holding more than 300 ml.

This way you will gradually stretch the time between going for a wee to a few hours. You get bladder control back and the bladder will no longer wrongly sense fullness at tiny volumes of urine. For most people, this is enough to return them to relatively normal bladder habits, full night sleeps and being able to toss out their maps of the cities marked with public toilets — and start looking at maps with landmarks and attractions!

Ignore what you may have been told as a child. Wait until you have a full bladder and avoid developing symptoms of bladder instability down the track. Cross your knees and hold those wees.

Prostate

In the male, the urethra passes through the prostate, a fibro-muscular gland.

The prostate manufactures fluids and squirts them into the seminal fluid during orgasm. For some unknown reason the prostate gland tends to enlarge during a man's lifetime. By the age of 70, 90% of men will have an enlarged prostate gland.

The symptoms of an enlarged prostate develop at the age of 65 in Whites, but at 60 in black people. These symptoms are – difficulty in starting urination, a reduction in size and force of the stream of urine, a sensation of incomplete emptying and dribbling after passing the water. About 10% of males will finally need some sort of surgery on their prostate gland.

Cancer of the prostate is the second most common cancer in men after lung cancer. And it is the third most common cause of cancer death in males aged over 55, after cancers of the lung and the colon.

Exercise

Yet another study has shown the benefits of exercise. (What a surprise!)

This study looked at 30 000 women aged between 54 and 79 years. The results were simple. Any exercise (even walking) reduced the incidence of urinary incontinence by about 25%. And the more physical activity they did, the greater the level of protection against urinary incontinence.

Why? Probably because any physical exercise made the muscles of the pelvic floor stronger.

References

Danforth, Kim N., et al., 'Physical activity and urinary incontinence among healthy, older women', *Obstetrics & Gynecology*, March 2007, pp 721–727.

Ganong, W.F., *Review of Medical Physiology*, 13th Edition, Connecticut: Appleton & Lange, 1987, p 605.

Gardiner E., Gray D.J. and O'Rahilly R., *Anatomy: A Regional Study of Human Structure*, 4th Edition, Philadelphia: WB Saunders, 1975, pp 460, 464, 1330.

Junqueira, L.C. and Carneiro, J., *Basic Histology*, 3rd Edition, California: Lange Medical Publications, 1980, p 407.

Rouse, Rada, 'Physical activity protects against weak bladder', *Medical Observer*, 23 March 2007, p 12.

Poisonous
Potatoes

One billion people worldwide regularly eat potatoes. In the USA alone the average consumption is about 60 kg per person per year. Potatoes taste good boiled, mashed, roasted, chipped, fried, in salads or as a packet of crisps (the salt helps). It's a lovely comfort food. Therefore, it is perfectly reasonable to think of the potato as a nutritious and harmless essential part of your diet. But the humble spud can kill.

Because plants don't have legs, they can't run away from Bad Guys. Instead, they have evolved whole galaxies of exotic chemicals to fight off attackers, e.g. insects, predators and various diseases. For this reason botanists are not surprised that potatoes can kill.

Botany of the Potato

Potatoes belong to the botanical family Solanaceae. It seems that this family name comes from the Latin word *solamen*, which means 'quieting'. The 'quieting' part of the name comes from the fact that the properties of some plants of the Solanaceae family could make you very quiet indeed. The plant mandrake could knock you out temporarily while the deadly nightshade could do it permanently.

The 2400 members of this family include tobacco, tomato, capsicum, eggplant, jimson weed, mandrake and deadly nightshade. Deadly nightshade is also known as *belladonna*, which means 'beautiful woman' in Italian. It was so named because it contained a chemical called atropine which, in very small doses, made a woman's pupils bigger — supposedly making her more beautiful. (The brains of human beings are wired to think that big eyes and small noses are beautiful —compare the heroes and villains in Disney movies.)

In the Middle Ages, some of the more deadly members of this family were thought to 'give witches the power to fly'.

Solanine - What It Is

The potentially poisonous chemical in potatoes is solanine. It was discovered in berries by a French chemist, M. Desfosses, in 1820, and found in potatoes in 1826 by another French chemist, M. Raup. Solanine protects the plant against pests and diseases, including the Colorado Beetle and the Potato Leafhopper.

Solanine was originally thought to be one single chemical. We now know that solanine is a mixture of many related chemicals, all going under the fancy name of 'potato glycoalkaloids'. However, the really nasty ones are two very closely related chemicals — a-solanine $\{C_{45}H_{73}NO_{14}\}$ and a-chaconine $\{C_{45}H_{73}NO_{15}\}$.

Solanine is produced in all parts of the potato plant, firstly in the leaves (which makes sense in terms of fighting off attackers) and then in the unripe fruits and flowers. In a potato, most of the chemical is found in the outer 3 mm of the flesh. There are many factors which affect how much solanine there is in the potato — the variety of potato; its maturity (higher in younger tubers); the intensity and colour of the light; stress (bruising, cutting and slicing); the temperature (warmer is worse); growing conditions (including the soil); and, of course, exposure to attackers.

Without solanine, potatoes taste very bland — but too much can kill.

Beware of the killer spud!

Are you looking at me?

The potato tastes good whether boiled, mashed, roasted, chipped, fried, in salads or as a packet of chips.
BUT, the humble spud is capable of killing!

A potentially poisonous chemical in potatoes is solanine.

Without solanine, potatoes taste very bland, but too much can kill.

It attacks the central nervous system interfering with how nerves talk to each other. Symptoms can include sweating, diarrhoea, bronchospasm, paralysis, heart failure, blindness, restlessness, confusion, weakness and coma (pretty much a bad situation whichever way you look at it).

The chemical structure of solanine

Solanine – What It Does

Solanine has two separate toxic effects.

First, it attacks the central nervous system by interfering with how nerves talk to each other. This pathway causes sweating, diarrhoea, broncho-spasm, paralysis, blindness, restlessness, confusion, weakness, heart failure and coma.

Second, it disrupts the movements of Ca^{++} (calcium ions) and Na^+ (sodium ions) across cell membranes, so interfering with generalised membrane absorption. This pathway leads to the bursting of red blood cells, excess fluid in body cavities and, in particular, problems in the gut (e.g. diarrhoea, vomiting and severe abdominal pain).

Solanine or Chlorophyll?

Even after it has been separated from its plant and taken out of the soil, the potato is still a living creature with its own metabolism.

The potato makes different chemicals, controlled very much by the quality of the light that lands on it — in the form of filtered and unfiltered sunlight or artificial lighting.

All artificial lights are different. There are fluorescent lights, tungsten lights, the special lights in the supermarket that are designed to make food more attractive, and so on. To the untrained eye all of these lights, especially if you have been under them for a while, simply appear white. However, each of these lights has its own colour balance — a reddish, greenish or bluish tinge, or yellow if it's a sodium light.

A reddish light landing on the potato will make chlorophyll, giving the potato a green colour. If the potato is hit by bluish or ultraviolet light, it will make solanine via a completely separate chemical pathway. However, in this case, the potato itself will not change colour. The two processes can happen at the same time, or not.

There are four possibilities for a potato. It can be green and rich in solanine or green and low in solanine. Or it can be non-green and rich in solanine or non-green and low in solanine. So what is the average person, without access to a full biochemistry laboratory, supposed to do?

What to Do

First, if it's green, you can just cut off about 5 mm of the potato flesh from the outer layer. Or you can just dump the whole potato into the compost. And second, if it's not green you should abide by the taste test. If it tastes bitter, don't eat it.

And just in case you want to 'know the enemy' and familiarise yourself with the taste of solanine, first grab a potato with little shoots growing out of it. Then, to taste solanine safely, just nip your teeth gently on the green tip of a shoot, taste it on your tongue and spit it out.

Solanine Has Killed

Farmers have long known that green or sprouted potatoes can kill stock. Actually, livestock are relatively resistant to solanine while human beings are very sensitive to it.

Over the years solanine in potatoes has killed about 30 people and made more than 2000 very sick. For example, a 1933 outbreak in Cyprus poisoned about 60 people, of whom one died. In North Korea in 1952–1953, 382 people were affected by solanine, with 22 dying. In an English school in 1978, 78 students were poisoned. Luckily, three of them who were close to death survived. In this case, uneaten potatoes, recovered two days later from residues intended for pig swill, were found to contain astronomically high levels of solanine — 250 mg of solanine per gram of peeled potato.

In laboratory tests, eating potatoes with this level of solanine gave a two-part sensation — an immediate bitter taste lasting about one minute followed by a strong burning taste lasting about five minutes. The taste is felt along the sides of the tongue and the back of the mouth. However, regular ingredients used in cooking (e.g. salt, oils and flavourings) can easily mask this bitter taste.

At this stage we don't exactly know what the fatal oral dose of solanine for human beings, but it's probably around 3–6 mg of solanine per kilogram of body weight. And although there's not a lot of hard data available, it seems that some people are more susceptible to solanine than others.

Almost certainly, the real figures for potato solanine poisoning are much higher than those officially recorded — but most cases would be lumped together with gastroenteritis or general food poisoning.

Solanine – The Good Side

Living creatures are complicated, so it's not surprising that solanine can have both good and bad effects. It has been used as a feedstock in the pharmaceutical industry to make steroids.

It was also used to treat bronchitis, asthma and epilepsy. Laboratory tests have shown it to be partly effective against lymphoma and cancer of the cervix, liver and stomach. It has anti-allergy, anti-fever and anti-inflammatory effects, works against some bacteria and viruses and can help in vaginal candidiasis.

Bless the Potato

What should you do to protect yourself, bearing in mind that half a kilogram of fully green potato can make a 45-kg person quite sick? Cut away any green sections. If the potato is bitter, don't eat it. Avoid fried potato skins — a popular appetiser in some restaurants — if they have any green tinge.

But don't become too paranoid. The philosopher and explorer Alexander von Humboldt said that South America gave us a great curse, tobacco, and a great blessing, the potato.

References

Conner, Herbert W., 'Effect of light on solanine synthesis in the potato tuber', *Plant Physiology*, American Society of Plant Biology, 1937, 12, pp 79–98.

Friedman, Mendel, 'Anticarcinogenic effects of glycoalkaloids from potatoes against human cervical, liver, lymphoma, and stomach cancer cells', *Journal of Agricultural and Food Chemistry*, 2005, Vol 53, pp 6162–6169.

Friedman, Mendel, 'Potato glycoalkaloids and metabolites: roles in the plant and in the diet', *Journal of Agricultural and Food Chemistry*, 2006, Vol 54, pp 8655–8681.

McMillan, Mary and Thompson, J.C., 'An outbreak of suspected solanine poisoning in schoolboys: examination of criteria of solanine poisoning', Quarterly *Journal of Medicine*, April 1979, pp 227–243.

Morris, S.C. and Lee, T.H., 'The toxicity and teratogenicity of solanaceae glycoalkaloids, particularly those of the potato (solanum tuberosum): a review', *Food Technology in Australia*, March 1984, pp 118–124.

'A Salt' the Water to Make it Boil

People who cook pasta almost always add some salt to the water. If you ask them why, they will say that the salt is added either to improve the flavour or to make the water boil at a higher temperature so that the pasta cooks faster.

However, one thing is certain — salt has hardly any effect on making the pasta cook faster.

Physics of Boiling Water

Adding heat to water turns the liquid water into steam. As the steam expands it pushes against the surrounding atmosphere. There is a balance of the steam pushing out and the atmosphere pushing in. At sea level on a normal, fine day this balance happens at 100°C.

However, at high altitudes where the atmospheric pressure is low, water will boil at a lower temperature. While in the Himalayas, I watched our porters bring water to the boil and add spaghetti — but it never fully cooked. We were about 3700 m above sea level. At this altitude the boiling point of water is only 88°C. No matter how long the porters boiled the water, it would never get hotter than 88°C, always leaving the pasta a little crunchy.

The opposite happens inside a pressure cooker. Because the pressure is significantly higher than atmospheric pressure, the water won't boil until it reaches 120–130°C. This higher temperature really speeds up the cooking.

Some people claim that adding salt to water does the same, i.e. it supposedly increases the temperature at which water boils. If the temperature is higher, then there is more energy in the system. So by adding salt to the cooking water the pasta will supposedly cook faster — significantly faster.

Pressure Cooker

The first pressure cooker was invented by the French physicist, Denis Papin, in 1679. It was called Papin's Digester, because it was so fast in cooking foods.

The pressure forces the boiling temperature of the water to 130°C. In this way the heat penetrates the food more quickly, preserving the vitamins and minerals while cooking the food more rapidly.

Water the Strange

Water, a very common but very unusual liquid, has kept physicists and chemists guessing for the past century and a half. This is a big achievement for such a simple chemical, which has only two atoms of hydrogen married to just one atom of oxygen.

First, compare H_2O to other similar liquids. The easy way to do this is to use the Periodic Table to find the elements that are similar to oxygen i.e. sulphur, selenium, tellurium and polonium, and then marry them to hydrogen. With added H_2 these chemicals — H_2S, H_2Se, H_2Te and H_2Po — have very low boiling points, the lighter ones having even lower boiling points. But H_2O bucks the trend. You would predict (from the graph on p 102) that its boiling point would be –70°C. Instead, it's +100°C — 170C° higher than expected.

Just add salt, it makes everything faster!

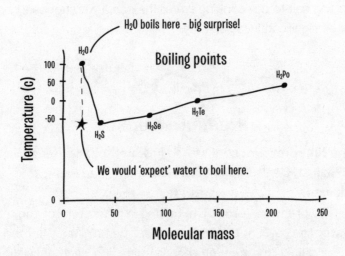

H₂O boils here - big surprise!

Boiling points

We would 'expect' water to boil here.

We've all been told that you are supposed to add salt to the water when cooking pasta, either to improve the flavour or to make the water boil at a higher temperature. However, one thing is for certain – salt has hardly any effect in making the pasta cook any faster.

And here's a second weird thing about water. As it cools, it becomes more dense (which you would expect) until it hits 4°C. Then it becomes less dense. This very unusual behaviour is still not fully explained.

And third, we still don't fully understand how water boils. The water molecule is shaped like a microscopic right-angled boomerang (actually the angle is closer to 104.5°). There's a negatively charged oxygen atom in the middle of the 'V' and two positively charged hydrogen atoms, one on each end of the boomerang's arms. In regular water, the H_2O molecules jostle around until the positive charge of one molecule attracts the negative charge of another. Therefore, at lower temperatures the water molecules are more 'sluggish', tending to 'stick' to each other. On average, one molecule of water will stick to four other water molecules. (Physicists say that water has 'short-range order', but no 'long-range order'). It's a fluctuating network, with the water molecules continually joining and breaking away from each other.

As you heat the water, you put more energy into the system, and these positive-negative attraction bonds start breaking up. They do not break up in an organised way — rather, they break up randomly. Surprisingly, it seems that as these attraction bonds break up, they create a microscopic cavity completely empty of water molecules. Smaller cavities merge to make bigger cavities which, in turn, somehow lead to molecules of water breaking free into the atmosphere. Yes folks, we've reached boiling point.

So Add the Salt

And yes, adding salt to water changes things. In pure water the water molecules are all 'fairly' organised. When you throw in some salt (sodium chloride), its atoms (as sodium and chloride ions) can wander around almost at random between the water molecules. These extra atoms increase the disorder, 'magically' increasing the temperature at which the water boils. (If you want to know more, look up any second year university Physics textbook on Phase Equilibria and the Clausius-Clapeyron equation — and while you're at it, read up on 'enthalpy'.)

So yes, salt increases the boiling temperature — but by the tiniest of tiny amounts. If you add 20 g of salt to 5 litres of water, instead of boiling at 100°C, it will boil at 100.04°C. A big spoon of salt in a pot of water will increase the boiling point by four-hundredths of a degree! So adding salt to water will not cook your pasta faster — unless your watch is calibrated in microseconds. (But the salt might still play havoc with your blood pressure.)

If you are in that much of a hurry you should run not walk with your pasta to the dining table. Salt doesn't significantly shorten the cooking time — even if you're a short-order cook. Just be happy with the knowledge that salt improves the flavour …

Different Sounds from Boiling Water

As you heat the water on the stove, you first hear a soft 'crackling' sound – the sound of tiny bubbles of air collapsing. The air is dissolved in the water, the increasing temperature driving the air from the liquid water to make tiny bubbles. These warm bubbles rise a short distance into the cooler water and then collapse. The collapse of each bubble sends out a sound wave, some of this sound escaping from the liquid into the air and then into your ear. The water is definitely heating and, just as definitely, these bubbles are made of air, not steam. (Steam is water when it's hot enough to be a gas.)

With further heating the sound gets louder. At this stage, you are now applying enough heat to the bottom of the pot to make microbubbles of steam. As the heat continues to be poured in, lots more bubbles are being made and getting closer to the surface before they collapse – so the sound gets louder.

Finally, the sound changes character and becomes softer, turning into a soft 'splashing' sound. At this stage, enough heat has been put into the system so that the bubbles make it all the way to the surface. This is called 'boiling'.

References

Ball, Philip, *H_2O: A Biography of Water*, London: Weidenfeld & Nicholson, 1999, pp 194, 195.

'Coalescing cavities bring water to boil', *New Scientist*, 4 December 2004.

Encyclopaedia Britannica, Ultimate Reference Suite DVD, 2006 — 'pressure cooker'.

Halliday, David, Resnick, Robert and Walker, Jearl, 'Boiling and the Leidenfrost Effect' from *Fundamentals of Physics*, John Wiley & Sons, 1993, Essay 6.

Sastry, Srikanth, 'Order and oddities', *Nature*, 18 January 2001, pp 300, 301.

Wolke, Robert L., *What Einstein Told His Cook: Kitchen Science Explained*, New York: W.W. Norton & Company, Inc., pp 46–48, 54, 62, 143–147, 167, 174–175.

Zahn, Dirk, 'How does water boil', *Physical Review Letters*, 20 November 2004, pp 227801–1 to 227801–4.

Is This Space Empty?

Amazingly, in the classic tag line from the famous sci-fi movie *Alien*, 'In Space, no-one can hear you scream', Hollywood got it right. In Space there are not enough molecules to carry the sound of screaming. But although most of us think that Space is totally empty of everything — it's not. There are lots of particles out in the so-called emptiness of Space. And although we have searched the depths of the Universe itself, we have never found a perfect vacuum.

History of Vacuum

The word 'vacuum' comes from the Latin word *vacuus*, meaning 'empty'. As far as scientists are concerned, a vacuum is pmerely a volume of space that contains no matter.

The first person to create a sustained vacuum was the Italian physicist and mathematician Evangelista Torricelli. In 1641, Torricelli was secretary and co-worker of Galileo during the last three months of his life. Two years later, in 1643, Torricelli carried out one of Galileo's suggestions. He took a glass tube about 1.2 m long, closed it off at one end, then filled it with mercury and tipped it upside down into a dish. Some, but not all, of the mercury flowed out — the space above the mercury in this tube becoming the first sustained vacuum created by a human being.

The height of the mercury measured approximately 760 mm. And even today we say that atmospheric pressure is equivalent to 760 mm of mercury. (Of course, there are other more modern units as well.) Torricelli noticed that the column of mercury changed its height from day to day and reasoned that this was due to changes in the pressure of the atmosphere. In his honour, one of the units used to measure atmospheric pressure is the 'Torr', equal to 1 mm of mercury.

The next person to work with vacuums was a German lawyer, physicist, engineer, natural philosopher and politician called Otto von Guericke. In 1650, he invented an air pump which made a vacuum. In 1654, he joined two copper bowls together to make a hollow

Pressure

We use the same 'quantity' to measure both a vacuum and the atmosphere. The usual method is 'pressure' – or the weight of something, divided by the area over which this weight is applied.

Air has weight. At standard room temperature (20°C) and air pressure, 1 m^3 of air weighs about 1.204 kg.

Let me give you a handle on this thing called 'air pressure'. Suppose you go outside, and mark out a square on the ground, 1 m by 1 m. The total weight of all the air above it extending all the way into Space would be about 10 tonnes. This means that the weight of the air pushing from the outside on each square metre of your windows is about 10 tonnes. The reason that the windows don't shatter is that this is exactly balanced by the same weight of air pushing from the inside.

Using technology, the lowest pressure or the highest vacuum ever reached is about one-billionth of one billionth of an atmosphere.

sphere about 35 cm in diameter — the famous 'Magdeburg Hemispheres'. He then removed most of the air from the hollow sphere with his air pump. These copper bowls were pushed together only by air pressure. He performed a series of famous experiments in front of Emperor Ferdinand III at Regensburg, in which two teams of eight horses could not pull the copper bowls apart.

Philosophy of a Vacuum

Ancient Greek philosophers were very worried by the concept of a vacuum. How could 'nothing' be something? Aristotle said, 'Nature abhors a vacuum'. Plato said that everything in the Universe was in some way related to an abstract Platonic ideal – and he couldn't imagine what the 'ideal' vacuum would look like.

About a millennium later, the Islamic philosopher Al-Farabi (850-970 AD) not only thought about the 'vacuum' but did experiments involving water and hand-held plungers.

A few centuries later, Christian thinkers began to worry about the idea of a vacuum. If a vacuum had nothing in it, then did this mean that there was no God in the vacuum? But didn't God have to be everywhere? In 1277, the French Bishop, Etienne Tempier 'solved' this problem by saying that God had no restrictions on His power and could therefore make a vacuum if He so wished.

Useful Vacuum

Surprisingly, 'nothing' is actually quite useful.

Vacuums are very important to the human race. We use vacuums in light bulbs, TV sets and electron tubes. We also use vacuums to put coatings on lenses to increase the transmission of

'In Space ... no one can hear you SCREAM'

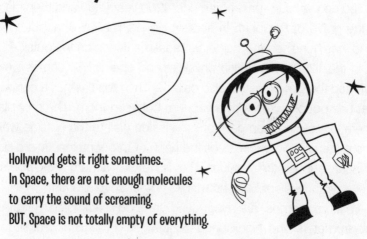

Hollywood gets it right sometimes.
In Space, there are not enough molecules
to carry the sound of screaming.
BUT, Space is not totally empty of everything.

The first person to create a sustained vacuum
was the Italian physicist and mathematician
Evangelista Torricelli way back in 1643.

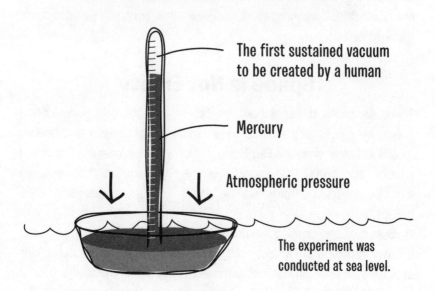

The first sustained vacuum
to be created by a human

Mercury

Atmospheric pressure

The experiment was
conducted at sea level.

light, to prepare plasma for blood banks and to work with metals, such as titanium, that would otherwise react with the atmosphere.

A vacuum gave us the world's oldest cake now on display in the *Alimentarium*, a food museum in Vevey, Switzerland. This cake was sealed and vacuum-packed around 4200 years ago and then placed in the grave of Pepionkh in ancient Egypt. The cake, about 11 cm wide, has honey on the inside and sesame seeds on the outside.

In the 1890s, the British physicist and chemist Sir James Dewar invented the Dewar vessel also called a Thermos flask — a container that has double walls with a vacuum between them. The only place where lots of heat can get from the inside wall to the outside wall is where the two walls join — at the neck of the container. It is difficult for heat to cross the vacuum. This is how a vacuum flask works to keep cold drinks cold and hot drinks hot for several hours.

Vacuum pumps are also used in food packaging, ultraviolet spectrometers and high-speed centrifuges. In fact, a centrifuge at the University of Virginia has a 14-kg rotor which is suspended in a vacuum by magnets. It started spinning at 1000 revolutions per second and is losing only about one revolution per second each day.

It was towards the end of the 19th century that scientists began to make pumps that could produce high vacuums. They wanted to study electrical discharges in gas, and this led to the invention of the light bulb.

Space is Not Empty

There are many different particles floating out there in Space once you leave the Earth's atmosphere. Some are hydrogen and helium atoms left over from the Big Bang. When the Universe cooled down enough about 400 000 years after the Big Bang, all the matter in the Universe was made up of about 75% hydrogen and 25% helium.

Some of the floating particles are bigger atoms and molecules thrown into Space by exploding stars. For example, when a star explodes smaller atoms get rammed together to make the atoms heavier than iron. This is virtually the only way that gold gets made.

All of the gold that you will ever touch was made inside an exploding star.

Besides hydrogen, helium and the stuff blasted out on the rare occasions when stars explode, there's also the 'solar wind' thrown out by normal stars. Our Sun (not a particularly large star) burns up 600 million tonnes of hydrogen every second, turning it into 596 million tonnes of helium every second. This 'missing mass' of 4 million tonnes per second gets turned into pure energy (via Einstein's famous $E=mc^2$ equation). The Sun blasts this energy out as heat, light, radio waves and so on. But besides throwing out enormous amounts of energy, the Sun also throws out the solar wind — about 200 000 tonnes of subatomic charged particles every second. (The solar wind is driven by this energy).

Think of our Sun as a dirty diesel truck, throwing out 'pollution' in all directions and creating a big pollution bubble around our solar system. When you factor in the more than 400 billion stars in our galaxy, you end up with a lot of stuff in so-called empty Space.

You can actually see the 'stuff' when a star explodes. When a star explodes as a super nova it blasts out a very powerful shock wave in all directions. This shock wave pushes an incredibly thin vacuum ahead of it, just like a road grader on a dirt road pushes the dirt in front of its blade. Since 1987, when a star exploded in a nearby galaxy, astronomers have watched the expanding circular shock wave grow bigger with each passing month.

More Than Particles

But wait, there's more than just particles in so-called 'empty Space'.

There are photons, neutrinos, Cosmic Background Radiation, the Missing Mass of the Universe (whatever this turns out to be) and even Dark Energy (whatever this turns out to be).

Measuring Nothing – Part 1

One way to get a feel for the emptiness of Space is to measure the distance between individual particles which range in size from large molecules, small molecules, atoms and ions to subatomic particles.

At sea level on Earth, air molecules are about *one-millionth of a millimetre* apart.

A few hundred kilometres above the ground where NASA's space shuttle orbits, the particles are about *one-hundredth of a millimetre* apart. These particles are a combination of an incredibly thin version of our Earth's atmosphere and the solar wind. There are enough particles to create a tiny 'wind drag' on spacecraft that orbit in this low earth orbit. In 1979, the first US space station, *Skylab*, fell to Earth over Australia thanks to this 'wind drag'. The ISS (International Space Station) also experiences this 'wind drag'. Because the station sinks closer to Earth by about 100 m or so every day, every few months it has to be boosted up again with rocket power.

Laboratory vacuum pumps here on Earth can do 10 times better, producing a vacuum where the particles are *one-tenth of a millimetre* apart. This is one of the reasons why scientists haven't really bothered trying to do vacuum experiments with the space shuttle. There's a further problem. In Space the orbiting molecules would crash into any experiment at around 27 000 kph!

Measuring Nothing – Part 2

NASA scientists have used the high speed of the space shuttle and a lump of metal to make a really high vacuum — better than anything we can make here on Earth. This device, called the Wake Shield, is basically a curved saucepan lid made of stainless steel, measuring approximately 4 m wide. The astronauts merely dump the Wake Shield in Space and then move the space shuttle a few kilometres away from it. As the Wake Shield zips through Space at 27 000 kph, it pushes atoms and molecules off to the sides. These soon sneak around to the back of the Wake Shield, but not instantaneously. This creates a high vacuum, where the atoms are about 1 mm apart, right behind the centre of the stainless steel

Physics of Vacuum

Scientists also have problems with the existence of a 'perfect' vacuum.

First, if this 'perfect' vacuum is inside a box then the walls of the box will give off radiation if it is at a temperature higher than Absolute Zero (−273.15°C or 0 Kelvin). If radiation is present then it's not a 'perfect' vacuum.

But can we get a 'perfect' vacuum if we get down to Absolute Zero?

Unfortunately, we cannot ever achieve the temperature of Absolute Zero because of the Heisenberg Uncertainty Principle which says that we can never exactly know the position of anything. At Absolute Zero everything stops and therefore, we cannot ever achieve Absolute Zero. However, in 2003, scientists at MIT (Massachusetts Institute of Technology) got very close – down to 0.45 of one-billionth of a degree above Absolute Zero.

Wake Shield. This is 10 times better than we can produce in our laboratories on Earth or 100 times better than you would normally find at this altitude. The Wake Shield has flown a few times and, in November 1996, was used to create a vacuum to grow some thin films of ultra pure, super conductor material.

Out in the solar system between the planets molecules, atoms and subatomic particles are about 1 cm apart — that's about the thickness of a finger. There are plans for spacecraft with enormous sails, kilometres wide, which could fly across the solar system powered by the solar wind.

And once you get out into the space between the solar systems, the particles are about 10 cm apart, roughly the distance across your clenched fist. This vacuum is pretty close to perfect nothing, but there are even better vacuums in Space.

Astronomers have just begun to look closely at other parts of the Universe, where the atoms are spread even further apart. We can't actually see these atoms directly because they don't give off lots of radiation. But off near the edge of the Universe are bright galaxies called quasars which shine light through this incredibly thin and almost perfect vacuum. Some of the light from these distant quasars is absorbed by these spread-out atoms. According to astronomers, these atoms and molecules are spread approximately 10 m apart. This is the 'thinnest' part of the Universe that we know of and at the moment, these are also the best vacuums that we know of.

Our Pollution Bubble

Interesting things happen when the particles of 'empty' Space and the solar wind collide.

Back in 1977, NASA launched the two *Voyager* spacecraft to visit the giant gas planets in our solar system. Between them, the spacecraft visited Jupiter, Saturn, Uranus and Neptune — and then kept on going way out of the solar system. They are still sending back data using their nuclear power packs and their microscopic 80 KB computer memories. By the way, 80 KB of memory is about 25 000 times smaller than the memory in my laptop.

Anyhow, it seems that about a third of a century after the two *Voyager* spacecraft were launched they are now getting close to where the pollution bubble of our Sun (the solar wind) hits the particles between the solar systems and creates a shock wave. Scientists will continue tracking the *Voyagers* over the next few decades, hoping that they keep transmitting data — and don't break down.

Big Shock Wave

There are locations in the Universe where the density is much greater, e.g. in some nebulae, which are interstellar clouds of dust, gas and plasma. They are the birthplaces of stars, and sometimes the places where stars die.

Of course, it's impossible with our current technology to get a photograph of our solar system's pollution bubble. But the Hubble Space Telescope took a photograph showing the pollution bubble from a star called L.L. Orion ramming into an especially dense part of Space — the Orion Nebula. This impact created a shock wave about half a light year across — a big event for supposedly 'empty' Space. Now that's something worth shouting about, not that anybody would hear you ...

References

Gugliotta, Guy, 'Voyager at edge of solar system', *Washington Post*, 25 May 2005.

Henbest, Nigel, 'Into the void', *New Scientist*, 25 April 1998, pp 26–30.

McGraw-Hill Multimedia Encyclopaedia of Science & Technology. Copyright © 1998 The McGraw-Hill Companies, Inc. — 'vacuum', 'vacuum pump' and 'Voyager'.

Bacteria vs Virus

I was spending a relaxing night watching junk television — one adult soap opera after another. But crammed in during an ad break was a short news update. It really cut into our light and frivolous night. The newsreader told her audience that there had been a disease outbreak in the Goulburn Valley area of Victoria, caused by the anthrax virus.

Her 'news' was about as accurate as the shows I'd been watching. Anthrax is caused by a bacterium *not* a virus. Understanding this difference can have a beneficial effect on your health. This ongoing confusion about the distinction between a virus and a bacterium leads to hundreds of parents every week asking their local doctor for antibiotics to cure their child's viral respiratory tract infection.

History of Anthrax

Anthrax is one of the oldest recorded diseases of animals. In the Bible in the book of *Exodus* 9: 8–9, it is written: 'Then the Lord said to Moses and Aaron, "Take handfuls of soot from a furnace and have Moses toss it into the air in the presence of Pharaoh. It will become fine dust over the whole land of Egypt, and festering boils will break out on men and animals throughout the land".' Ancient Greek and Roman authors such as Homer, Hippocrates, Ovid and Pliny also mention the disease.

The word 'anthrax' originally comes from the Greek word meaning 'coal', referring to the coal-black wounds that appear on

Bacteria vs Virus

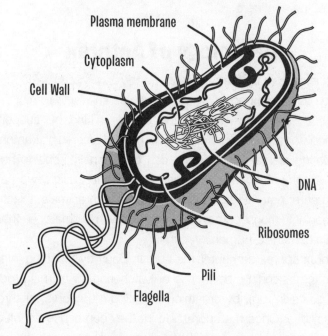

Plasma membrane

Cytoplasm

Cell Wall

DNA

Ribosomes

Pili

Flagella

generic bacterium
(lots of stuff necessary so that it can live and reproduce)

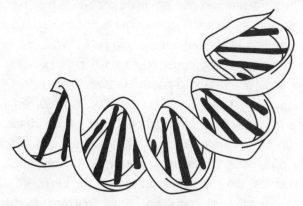

virus
(DNA and bit of protein – it has none of the machinery
needed to live and reproduce)

the skin. Even today, it infects 20 000–100 000 people worldwide each year. In Australia, the Anthrax Belt (where anthrax, although uncommon, is most likely to occur) is located in the southeastern corner of the continent.

Pathology of Anthrax

Anthrax is caused by the bacterium *Bacillus anthracis*.

When times are bad or when the bacterium wants to rest, it can exist as a spore, surviving in the soil in a kind of 'suspended animation' for a century or more. The natural reservoir of anthrax is grass-eating animals, although it can occasionally jump to human beings. Nearly all animals are susceptible to it. Luckily, we have a much greater natural resistance to anthrax than animals. Practically all human infections come from contact with animals — there is very little human-to-human infection.

Anthrax spores can enter the human body by touching infected animals (e.g. skinning, dissecting or butchering) or animal products (e.g. a goatskin rug), by breathing or eating the spores, or through an insect bite. Once the spores are in, they can cause an infection that has an incubation period of 1–7 days. The spores 'wake up' within hours, turn into living bacteria and multiply to incredible numbers. The bacteria then give off 'anthrax toxin' which has three main proteins — protective antigen, edema factor and lethal factor.

Anthrax can attack different parts of the body. When the spores enter through the skin they cause 'cutaneous anthrax'. This form accounts for 95% of cases and has a 10–20% death rate. Spores entering the lungs cause 'inhalation anthrax' (sometimes called wool-sorter's disease). This form accounts for 5% of cases, with close to a 100% death rate. Although rare, anthrax can attack the gut (with a 50% death rate) and the brain (usually fatal). The treatment involves the oldest antibiotic of all, penicillin. If the sufferers are sensitive to penicillin, other antibiotics such as ciprofloxacin, erythromycin, tetracycline or chloramphenicol are administered.

Bacteria

The name comes from the Greek word *bakterion*, the diminutive of *baktron* meaning 'cane' or 'staff'. These micro-organisms were so called, because the first ones to be discovered had the shape of tiny rods or canes.

The Gods must love bacteria, because there are so many of them in a wide range of differing environments on the planet, the vast majority of them harmless to human beings. They range in size from 0.1 to 12 microns (millionths of a metre) wide. You can't see them with the naked human eye — you need a light microscope to see them.

Bacteria have been around for over 3.5 billion years, making them the longest surviving creatures on Earth. They exist in the air we breathe, in the water we drink, in the food we eat — and yes, in the soil we walk on.

Bacteria are the smallest creatures that have all the machinery necessary to live and reproduce. Suppose you have a pool of water filled with all the food types that a bacterium would need to survive, i.e. fats, proteins and carbohydrates. If you place one single bacterium into this pool, it would eat and grow bigger, multiplying and filling the pool with billions and trillions of bacteria within days.

Luckily for us, the biochemistry of bacterial cells has a few significant differences from the biochemistry of human cells. Therefore, in the 1940s, it was relatively easy to develop drugs (antibiotics) that would kill bacterial cells but not harm human cells. This is why we have many hundreds of types of antibiotics today.

Antibiotics are the reason why most bacterial diseases are not the killers that they were in the past.

Viruses

Viruses, however, are something else.

The name 'virus' comes from the Latin word for 'slimy liquid' or 'poison'. The Gods probably love viruses too, because there are lots of them in the environment as well. They are much smaller than bacteria, ranging in size from 20 to 400 nm (billionths of a metre).

They are so small that they can only be seen under an electron microscope.

Because they are not living cells they do not have the machinery to live and reproduce. Instead, they exist mostly as a kind of formula to make a virus — just the bare minimum of DNA or RNA wrapped in a few proteins to make it look like a nice meal for a bacterium or other living cell. If a virus is lucky enough or cunning enough to get swallowed inside a living cell, it takes over, forcing the internal machinery of this cell to make lots more viruses. However, if you place one single virus into a pool filled with fats, proteins and carbohydrates — and no cells of any type whatsoever — it will never increase its numbers even after millions of years. Like parasites, viruses depend on the cell that they have invaded for practically all of their functions, including making more viruses.

Because viruses can only do their stuff inside cells, any drugs that have an effect on them will usually attack our human cells first. We have been developing drugs that attack viruses for only a few decades and, even today, we still have only a handful.

So let me give it to you straight. Antibiotics will attack bacteria but are useless against viruses. There is absolutely no point in taking antibiotics for a viral infection.

Side Effects

All drugs have side effects. Paracelsus (the 16th century physician) said, 'All drugs are poisons, what matters is the dose'.

The general public seems to believe that the side effects of antibiotics are the worst of any class of drug. However, their side effects are quite minor compared to the good that they do.

If you want drugs with serious side effects, try some of the anti-cancer or heart disease drugs.

The Bottom Line

Anthrax was the very first disease to be related to a specific bacterium — discovered in 1863 by the French biologist Casimir-Joseph Davaine. It was also the very first disease to have a vaccine developed against it — achieved by Louis Pasteur in 1881.

And what happened in the 2007 Goulburn Valley anthrax outbreak in Victoria? A few dozen animals died before the disease could be contained, many more were vaccinated against anthrax and didn't die — and there was no loss of human life. Public health measures swung into action and the anthrax came and went, with no danger of a major outbreak in either human beings or animals.

Our efforts to fight bacteria have been continuing for a long time. Even so, a century and a half later, ill-informed journalists still confuse viruses and bacteria — and help spread the confusion that sees patients pleading with doctors for antibiotics to treat a virus.

References

Chin, James, 'Anthrax' in *Control of Communicable Diseases Manual*, 17th Edition, Washington DC: American Public Health Association, 2000, pp 20–25.

Encyclopaedia Britannica, Ultimate Reference Suite DVD, 2006 — 'anthrax', 'bacteria' and 'virus'.

Hansen, Thomas M., 'Correlation between mechanical strength of messenger RNA pseudoknots and ribosomal frameshifting', *Proceedings of the National Academy of Sciences*, 3 April 2008, pp 5830–5835.

Wilson, J.D., et al. (Eds), 'Anthrax' in *Harrison's Principles of Internal Medicine*, 12th edition, New York: McGraw Hill, Inc., 1991, pp 575–577.

Mashed Facts

The potato is a fine vegetable, high in nutrients and low in fat. In fact, in French (*pomme de terre*), Dutch (*aardappel*) and Austrian German (*erdapfel*), it is known as the 'apple of the earth'. The fourth most important food crop in the world (after rice, corn and wheat), the potato has an advantage over the top three crops in that it can be grown more quickly, in harsher conditions and on less land — and still deliver more produce per hectare.

However, when things go wrong, as in the Great Potato Famine of Ireland from 1845 to 1849, starvation can follow. Ireland and the potato are symbolically linked — but the potato did not originally come from the Emerald Isle.

History of the Potato

The potato actually came from South America, having been cultivated there for about 8000 years. In fact, the Peruvians have been freeze-drying potatoes for over 2000 years. There are approximately 13 000 different varieties of potato, with only 8000 in common use. When the greedy Spanish Conquistadors invaded South America in the early 1500s, they were looking for gold and silver, ignoring the humble potato that they trampled under their feet. But today's annual potato crop of approximately 300 million tonnes is worth three times all of the precious metals that the Conquistadors stole. The Inca empire was crushed and the Spanish empire has in turn faded away, but King Potato still reigns supreme. There was some resistance to this new vegetable, when it was introduced into Europe. For example, the Scots shunned the potato because it was not specifically mentioned in the Bible. However, the potato had some undeniable advantages over other crops. Not only can the potato be grown at sea level and up to an altitude of 4.2 km, but it will also produce 4–5 times the yield of rice and wheat. For this reason the potato was eventually accepted.

There was even the Potato War in 1788. The Prussians and the Austrians fought each other until their armies ran out of potatoes to feed the soldiers. The soldiers then simply walked away from the battlefield, leaving it deserted.

Potatoes eventually became essential to Europe. In fact, the Industrial Revolution could not have happened without the potato — a cheap and hardy source of food for the new workers who had been funnelled into the cities.

Potatoes and Ireland

Sir Walter Rayleigh introduced the potato to Ireland. And indeed, Ireland was the first country to have the potato as its main vegetable crop.

T'be sure, t'be sure ... the potato didn't originally come from Ireland

The humble potato

In terms of genetic diversity, Ireland did not get several thousand varieties of potato nor even several hundred — it got only two varieties. Unfortunately, these were both very susceptible to a disease of potatoes called 'late blight'. Late blight is caused, not by a fungus (another myth) but by the organism *Phytophthora infestans*, also known as an 'oomycete' or 'water mould'. The organism is definitely not a fungus. It is more closely related to kelp than mushrooms.

Late blight devastated the Irish potato crops from 1845 to 1849. At the time the Reverend Miles Berkeley identified this potato disease as being caused by an infectious agent. In this matter he disagreed violently with his religious colleagues who blamed the devil instead.

The destruction of the potato crops dumped a toll on the Irish population and culture that was shocking and huge. It was caused by many different factors, including the actual shortage of food, the unsympathetic British military occupation, and a bizarre social structure involving absentee landlords and tiny plots of land.

During the Great Potato Famine, one million people starved to death and another million emigrated overseas. The Great Potato Famine helped cause the demise of the Celtic Church and the near-annihilation of the Irish language. The population plummeted from its previous level of 8.4 million. It hadn't even returned to half of this number by the time of Irish Independence in 1921, nearly three-quarters of a century later.

Genetic Diversity

In Australia, we have relatively easy access to about 30 varieties of potato. Because they have a reasonable degree of genetic diversity, at least some of them should be resistant to a nasty disease. However, in Ireland in 1845, both varieties of potato were frighteningly susceptible to *Phytophthora infestans*. The devastation of Ireland, combined with the fact that Ireland was the first country to have the potato as its staple food crop, forever associated Ireland with the potato. This is why many people believe that the potato came from Ireland.

While Irish people and Irish culture have given us many things, it did not include the first cultivation of the humble spud.

Spud

The name 'spud' comes from the small spade used to dig potatoes out of the ground.

Industrial Peeling and Chipping

People enjoy eating hot chips but don't want to peel and chip them. So the fast-food industry came up with an industrial-strength method of getting the skin off – and it's not very pretty.

Peeling is the first stage in making chips, which is simply a matter of exploding off the skin. The potato is put into a giant pressure cooker and boiled at a pressure of 15 atmospheres. Any moisture under the skin of the potato would love to expand and turn into steam, but the 15 atmospheres of external pressure does not allow this to happen. But when the external pressure is abruptly released, the water under the skin expands and the skin explodes off the potato like a hand grenade.

And how do they turn this naked, shell-shocked potato into chips? Well, they put it into a pipe with water flowing through it at 80 kph. The potato then commits hari-kiri, plunging itself onto a set of knife blades and turning itself into a bunch of skinny chips.

References

Money, Nicholas P., 'Reverend Berkeley's devil', *Nature*, 7 June 2001, pp 644–645.

Nickens, T. Edward, 'Sherlock of spuds', *Smithsonian*, December 2002, pp 30–31.

O'Neill, Gearoid, 'Parasite landlords left Irish to die in famine', *The Daily Telegraph Mirror*, 22 August 1994, p 28.

Pearce, Fred, 'The famine fungus', *New Scientist*, 26 April 1997, pp 32–36.

Rhoades, Robert E., 'The incredible potato', *National Geographic*, May 1982, pp 668–694.

Ristano, Jean B., et al., 'PCR amplification of the Irish potato famine pathogen from historic specimens', *Nature*, 7 June 2001, pp 695–697.

Sattaur, Omar, 'Native is beautiful', *New Scientist*, 2 June 1988, pp 54–57.

'This spud's for you', *Science Digest*, December 1988, pp 26, 109.

Mellow Yellow

Australian scientist Sir Gus Nossal has said that because the problem of radioactive waste already exists, we must address the issue properly and find a permanent solution to disposing of it safely. Permanent solutions are better than fanciful ones, e.g. allowing tectonic plates to carry our nuclear waste deep under continents, shooting it into the Sun or sinking it into the polar ice.

Nuclear waste is not a relatively recent development — it's been around for billions of years. In fact, there were natural nuclear reactors on Earth nearly two billion years ago, and their radioactive wastes have remained safely buried until now.

Meet Uranium

Uranium (symbol U) is a soft, white metal, about 19 times more dense than water. It was first discovered in 1789 by the German chemist M.H. Klaproth, who named it in honour of the planet Uranus, discovered only eight years earlier.

Uranium burns easily at 170°C, but if you grind it up very finely, it will burst into flames as soon as it is exposed to air.

Natural uranium is a mixture of three radioactive isotopes or varieties of uranium. They are almost identical chemically but each has a different weight.

No matter where you dig it out of the ground, the ratio of these different isotopes is the same everywhere on Earth. The ratio is approximately 99.28% uranium 238 (U-238), 0.71% uranium 235 (U-235) and 0.006% uranium 234 (U-234). Tinkering with these percentages gives you such terms as 'enriched uranium', 'depleted uranium' and 'weapons-grade uranium'.

In other words, if you processed some uranium ore from the ground and extracted 100 g of pure, 100% uranium from it, 99.29 g will be the other two natural isotopes of uranium. The remaining 0.71 g will be U-235. This is what is meant by the 'level' or 'concentration' of U-235.

In technical terms U-235 is the only isotope of uranium that goes 'bang'. In an enriched concentration it is the only isotope that can be used in nuclear reactors and in nuclear bombs.

Natural nuclear reactors cannot exist today, because the present, natural levels of U-235 are much too low. This is why it took the combined brain power of a bunch of Nobel Prize-winning physicists, led by Enrico Fermi, to design, build and run the first artificial nuclear reactor in what had been a squash court at the University of Chicago on 2 December 1942.

Two Billion Years Ago

But two billion years ago, when the Universe was a lot younger, the level of U-235 was about 3% — easily high enough to be used in today's nuclear reactors. And it was then that the story of the natural nuclear reactor began.

A special set of climatic and geological conditions brought together many thousands of tonnes of uranium in an underground bed. The uranium, in seams 5–10 m thick and 600–900 m wide, was located in what is now Oklo, in Gabon, West Africa.

1 – U into River Beds

Originally, the uranium was deposited in rocks over a very large area of approximately 35 000 km^2. Over hundreds of millions of years, rain eroded the rocks. The water washed the uranium down from the hills dumping it into the rivers, in the same way that other heavy elements like gold end up in rivers today. In this first stage, the uranium was concentrated into the river beds. At this time, there was very little oxygen in the atmosphere.

Shizzle my yellow fizzle

Sir Gus Nossal* – Australian scientist –
has said that the problem of radioactive
waste already exists, so we have to address the issue properly
– and find a permanent solution.

*(Sir Gus is actually much better looking)

Africa

Gabon Oklo

The uranium isotopes found at Oklo
strongly resemble those in the spent
nuclear waste generated by today's
nuclear power plants.

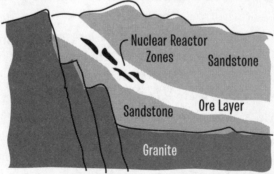

Nuclear Reactor
Zones

Sandstone

Sandstone

Ore Layer

Granite

The radioactive remains of a natural nuclear fission reaction
that happened 1.7 billion years ago in Gabon, Africa,
were held in place by the surrounding geology.

Luck

It took quite a few coincidences for the natural nuclear reactors at Oklo to exist.

First, you need the appropriate geological and biological accidents to bring the uranium together in one spot, and then concentrate it even further.

Second, according to physicists, you need the percentage of U-235 to be at least 1%. This means that natural nuclear reactors could have operated at any time until about 400 million years ago. In fact when the Earth formed about 17% of natural uranium was U-235. Two billion years ago, it would have been ridiculously easy for a scientist to build a nuclear reactor. But it was harder for nature to build one.

Third, you need the concentrated ore to be in seams at least two-thirds of a metre thick. If the seams are any thinner, too many neutrons would escape and the nuclear reaction would not be self-sustaining.

Fourth, you need something like water around the uranium. As uranium atoms split (which they do all the time) they give off neutrons. The neutrons have to be slowed down. Normally, they fly away with a very high energy, and end up becoming absorbed by U-238. However, if you slow them down with hydrogen (which is part of water, H_2O), they are much more likely to be absorbed by U-235.

Finally, a nuclear reactor cannot work if there are large quantities of elements that absorb neutrons. Nuclear physicists call these elements (e.g. lithium and boron) 'poisons'. Luckily, there were no 'poisons' in the earth at Oklo.

2 – Dissolve U with Oxygen

The second stage began about two billion years ago with the appearance of blue-green algae — the first creatures able to carry out photosynthesis, i.e. they are able to get energy from sunlight. This gave them a tremendous biological advantage. While every other creature had to find and eat food to get energy, all the blue-green algae had to do was float around and soak up the sunlight. This algae multiplied and began to release huge amounts of a very corrosive by-product into the atmosphere — oxygen.

This oxygen dissolved in the waters, oxidising the uranium — and, as it turns out, oxidised uranium is quite soluble. Therefore, the uranium that had been lying in little pockets on the bottom of the creeks and rivers dissolved into the water, the natural flow of the rivers carrying it downstream.

Half-life

The half-life of a radioactive element is the time needed for half of the original atoms to spontaneously decay.

For example, if you start with two million atoms of U-234, it will take 247 000 years for them to decay to just one million atoms of U-234. The half-life for U-235 is 713 million years, while the half-life for U-238 is longer again at 4.51 billion years.

So two billion years ago, when the Earth was 60% of its current age and the Universe was 85% of its current age, there was a lot more U-235 in the Universe because it hadn't yet decayed. Natural uranium would have had about 3% U-235 – enough to fuel a nuclear reactor.

3 – Concentrate U

The uranium floated in solution until it reached the delta of the river system. Because the organic ooze at the bottom of the river delta was very low in oxygen, the waters immediately above the ooze were also very low in oxygen.

The uranium deoxidised, became less soluble and fell out of solution into the ooze at the delta mouth. The uranium that had originally covered 35 000 km^2 was now concentrated at Oklo into an area of a few square kilometres.

4 – Cover, Add Water

Over millions more years, the river carried down more rocks and sand, dumping them on top of the uranium. The uranium ore was squashed by the weight of these upper layers to make a layer of radioactive sandstone. At this stage, the uranium ore was now 50–500 m underground.

Then the earth moved and the sandstone cracked. As it cracked, the uranium ore body fractured and water began to trickle from above through 50–500 m of 'overburden' down to the ore. The water did two very important things.

First, the water dissolved some of the uranium and carried it further underground. As the water dried out, it left behind small pockets of very concentrated ore. These pockets were about the size of a back-yard swimming pool.

5 – Natural Nuclear Reactor

Second, the water reacted with the neutrons coming off the U-238 as it decayed. It modified these neutrons so that they were able to split the atoms of the U-235. It turned them from 'fast' neutrons to 'slow' neutrons.

The so-called 'fast' neutrons cannot make U-235 go bang. Only so-called 'slow' neutrons can do this.

When Uranium Splits ...

When an atom of U-235 absorbs a neutron, the nucleus gets excited, changing its shape from round to egg-shaped. It will return to its original shape 15% of the time. But 85% of the time, it will continue to change shape until it looks like a peanut shell. The nucleus will then give off either 2 or 3 neutrons and split into two unequal fragments. These fragments are radioactive and, in time, will themselves decay further. Eventually, the single atom of U-235 will give rise to about 30 different stable daughter elements.

At Oklo, more than half of these daughter elements are still present in the ore body. The only ones missing are those that dissolve in water, and the gases.

Lo and behold, almost two billion years ago, the first nuclear reactor on Earth fired up, breaking the U-235 into smaller atoms and giving off heat. At the same time, it also created some radioactive waste. About 30 minutes later, the reactor was so hot that the water turned into steam and escaped, pushing its way through the soil. No water meant no slow neutrons and the reactor switched off. After a few hours, the reactor had cooled down enough to allow more liquid water in, this providing more slow neutrons. And so it started up again.

The reactor ran like a geyser in a volcanic park — on for half an hour and off for 2–3 hours. It stuttered along like this for a few hundred thousand years, until the U-235 was 'burnt' to make waste, about 50–500 m underground. Until it was mined, the uranium had moved fewer than 4 m from where it was made nearly two billion years ago.

So far, we have discovered about 18 of these burnt-out natural nuclear reactors in Oklo. They were not particularly efficient, each one generating an average power of less than 100 kW — enough to run a few dozen toasters.

How Do We Know?

The detective story began in the early 1970s when a nuclear fuel processing plant in France was sent uranium from a mine in Oklo.

Strangely, the uranium ore had slightly lower than normal levels of U-235. By itself this was very unusual but only a very careful technician would have noticed. Later, however, technicians found some samples that had much less than the normal level — 0.44% U-235 (instead of the standard 0.71%)!

In a nuclear reactor, U-235 'burns' and turns into two other elements. So the technicians went looking for some of these elements that are the by-products of a nuclear fission, U-235 reaction. They found them in the unprocessed uranium ore. The conclusion was obvious, but unbelievable. There had been a nuclear power plant at Oklo, in Gabon, nearly two billion years ago.

Two billion years (that's 2000 million years) is an immense period of time. Animal life left the oceans to invade the land only 400 million years ago and human beings have been around for no more than 3–5 million years.

Problem

Since the 1960s, approximately 400 nuclear reactors in about 30 countries have produced around 200 000 tonnes of radioactive wastes. Each year, another 10 000 tonnes are added. The wastes are usually stored above ground on the site where they are generated, definitely not a permanent solution.

Today we are stuck with several hundred thousand tonnes of radioactive wastes. For better or for worse, well-meaning political action has prevented plans for their long-term safe disposal from occurring.

From Curiosity to Nuclear Reactor

Not regarded as dangerous, uranium was merely a colouring agent and a laboratory curiosity until the 1920s.

Uranium salts were used to give glass and pottery glazes (including household crockery) a yellow colour that was still being done until the 1960s. If you have a pair of World War II binoculars with yellow-tinted lenses, the yellow colour is probably due to uranium.

Then came the discovery of radium, a highly radioactive element extracted from uranium ore. It was assumed to have health benefits and was even added to tonics!

Radium also became the miracle cure for cancers. So radium was extracted from the uranium and the tailings simply thrown away or stockpiled. But during World War II, these old tailings suddenly became very valuable. All three isotopes of uranium are radioactive, but the military were after just one of them – U-235. It was used in the first nuclear bomb detonated over Hiroshima in Japan on 6 August 1945, killing more than 75 000 people.

U-235 is also the principal fuel used for the generation of electricity by nuclear reactors.

Solution?

The 18 natural nuclear reactor sites of Oklo have given us a good example of how nature itself kept radioactive wastes safely buried for two billion years.

Unfortunately, digging a hole a few kilometres into the ground to bury nuclear wastes is too expensive for most governments. It is cheaper to leave nuclear waste in steel drums in a shallow, plastic-lined ditch. However, in Switzerland, where 40% of their electricity requirements comes from five nuclear power plants, they have 'bitten the bullet'. They are taking 40 years and spending $3 billion to bury their nuclear wastes 1.2 km underground — 5% of the cost of generating the electricity.

It sounds easy. Just dig a hole a few kilometres deep in a geologically stable area, far away from any water table or aquifer. Then, at the bottom of the hole, dig out a large number of horizontal side-tunnels — all radiating from the central hole like the spokes of a bicycle wheel —and fill these tunnels with radioactive wastes. Then come up 20 m and fill the central tunnel with 20 m of concrete. Keep repeating the entire process until you get to within a kilometre of the surface, then stop digging side-tunnels and fill the hole with 1 km of concrete.

Yes, this would be very expensive and, at the same time, very safe.

References

Chapman, Neil and McKinley, Ian, 'Radioactive waste: back to the future?', *New Scientist*, 5 May 1990, pp 36–40.

Cowan, George A., 'A natural fission reactor', *Scientific American*, July 1976, Vol 235, No 1, pp 36–47.

de Laeter, J.R., 'The Oklo reactors: natural analogues to nuclear waste repositories' *Search*, Vol 16, No 7–8, August/September 1985, pp 193–196.

Meshik, Alex, 'The workings of an ancient nuclear reactor', *Scientific American*, November 2005, pp 82–91.

Weiss, Peter, 'Primordial nukes: the 2-billion-year-old tale of Earth's natural nuclear reactors', *Science News*, 12 March 2005, pp 170–172.

Oxymoron

About once every decade, the Oxygen Healing myth recirculates. Loosely speaking, it claims that diseases are caused by low levels of oxygen in the body. So, in order to counteract the low oxygen, you have to buy brand X or Y's oxygen product or undergo oxygen therapy to be guaranteed perfect health. These oxygen treatments go under various names from the snappy 'Oxyrich', 'Liquid O2' and 'Vitamin O' to the alarmingly scientific 'Hyperoxygenation', 'Oxidology', 'Oxymedicine' and 'Bio-oxidative Therapy'.

Unfortunately for the sufferers of the various diseases, oxygen therapy doesn't cure any of them. (Of course, this doesn't include the few heart and lung illnesses where the sufferers need to inhale extra oxygen just to stay alive.)

Crazy Theories – General

All the theories behind curative oxygen therapies are based on the simple premise that if a little oxygen is good for you then more has to be better. And oxygen healing is promoted as 'natural and nontoxic'.

The underlying basis for oxygen therapy is that all diseases are caused by a lack of oxygen in the body. This has never been proved! In fact, the theories behind oxygen therapy are so weak as to be laughable.

Numbers – Oxygen Used

The heart of the average 70-kg person will pump approximately 250 ml of oxygen each minute to be used by the bodily tissues. That's 360 000 ml of oxygen each day.

Typical oxygen 'healing' treatments will deliver 100–1000 ml of oxygen – miniscule compared to the 360 000 ml that you naturally supply and use each day.

Crazy Theories – Specific

In 1924, Dr Otto Warburg discovered that cancer cells used an inefficient method of burning oxygen. Of course, to the oxygen healers this implied that the way to kill cancer cells was to flood them with oxygen.

However, the opposite is true. Cancer cells grow very rapidly and need all the oxygen that they can get. Therefore, most cancers have, as part of their mode of survival, the ability to force the body to grow more blood vessels to feed the cancer with lots of oxygen. In fact, some cancers are treated by killing the blood vessels and thus starving the cancer of oxygen.

Dr Warburg won the Nobel Prize for Physiology or Medicine in 1931 for his work on the 'discovery of the nature and mode of action of the respiratory enzymes'. (The people who promote the oxygen-heals-everything theory claim that he won two Nobel Prizes, but he didn't.) Later research showed that both cancer cells and normal cells used the same methods of burning oxygen.

A second theory suggests that low levels of oxygen in the body lead to a build-up of toxins. This has also never been proved. Over the past decade, our advancing technology has given us the ability to detect absolutely microscopic amounts of various chemicals — but we have never found these postulated toxins.

Today's snake oil

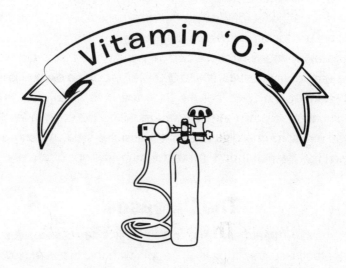

Breathe easy, because just one 'hit' of 'Liquid O₂' could cure whatever you might have, think you might have ... or plan on getting!

In 1919, a Detroit doctor, William F. Koch, came up with his theory on Oxygen Healing. He claimed that the root causes of all human diseases were toxins in the body, resulting from all of the injuries sustained over one's lifetime. Luckily, these toxins could all be destroyed by his wonder oxygen chemical, 'glyoxylide'. It had a very simple chemical formula — $O = C = C = O$. Not only has this chemical never been found, studies in theoretical chemistry have proved that it cannot be made.

A separate theory claims that all our woes are caused by bacteria. It then asserts that all bacteria evolved in the past when the levels of oxygen in the atmosphere were much lower. So, if the bacteria happen to find a spot in the body where the oxygen levels are low, they will multiply and cause us harm. Both of these assertions are false!

Yet another theory claims — again, with absolutely no evidence to support it — that low levels of oxygen in our tissues will make our bodies more acidic. This theory then claims that this mysterious 'acidity' would deplete our 'vital energy' (which, by the way, is a concept unknown to modern physiology). This is incorrect. Our kidneys have the job of keeping our acid-base balance finely tuned, and if we got too acidic, we would die.

And just to round off these crazy theories, one theory claims that in the very recent past (5000–10 000 years ago) the oxygen level on Earth was around 30% (not the 21% that it is today). But human beings chopped down all the trees, reducing the oxygen level and so causing all of the diseases that we suffer today. Wrong again! Oxygen has been at the 21% level for many millions of years.

The Diseases

If you have a 'treatment' that 'cures' only one disease, you limit your potential customer base to people with this disease. However, if your 'treatment' cures everything from sunstroke to syphilis, then you have a much larger group of sufferers to take advantage of.

By an amazing coincidence, oxygen therapy seems to be able to cure every disease known to humanity, including migraine, cold, flu, deafness, high, low and irregular blood pressure, as well as skin rash, AIDS, earache, softening of the brain, Down's Syndrome, gum disease, Alzheimer's and Parkinson's Disease, *and* let's not forget herpes, hepatitis, chronic fatigue syndrome, cytomegalovirus, arthritis, systemic candidiasis, ageing, all cancers and — you guessed it — loss of sexual function.

The Cure? H_2O_2 (Introduction)

Today, there are many methods offered to get this unnecessary extra oxygen into your body. Practically all of them have no benefit, but they can have side effects ranging from mild discomfort to death.

In 1888, Dr I.N. Love published his work in the *Journal of the American Medical Association*, claiming that applications of hydrogen peroxide (H_2O_2) could cure cancer of the uterus, diphtheria and scarlet fever. His claims have never been proved. But hydrogen peroxide has become the most popular way of adding oxygen to the body.

Other Reasons Why You are Low in Oxygen

Here are three ridiculous reasons as to why we supposedly need extra oxygen:

- The weather and pollutants in the atmosphere chew up all the available oxygen. (Wrong! The oxygen levels are very constant.)
- Our nutrient-depleted soils make the food that we eat low in oxygen-carrying capability. (Wrong again! Food doesn't deliver oxygen to us, the air does.)
- Eating 'bad' foods overloads the body with 'toxins and impurities', which make the blood unable to absorb oxygen from the lungs and carry it to where it's needed. (And wrong again! Cyanide is one of the very few chemicals that directly interferes with oxygen metabolism.)

And here are three crazy reasons that have *never* been proved.

- The stresses of the world give us emotional imbalance and unhealthy mental attitudes that rob the body of oxygen.
- Infections and cancers rob our bloodstream of all its oxygen, so Oxygen Healing Therapy is needed to 'counteract their venom'.
- Fluoridation of water interferes with the metabolism of the mitochondria, thereby interfering with oxygen metabolism.

Hydrogen peroxide (H_2O_2) is a clear liquid. It is similar to water (H_2O) but has an added oxygen atom. Once out of the bottle, it rapidly breaks down because it is unstable. One litre of 30% H_2O_2 releases 100 litres of gaseous oxygen (at 3% H_2O_2, it will release 10 litres).

There is no doubt that hydrogen peroxide (3% solution in water) can clean your wooden cutting board or your driveway, bleach your hair, and disinfect your toothbrushes. But the liquid is not very good at penetrating human flesh and has only weak germicidal properties.

Liquid hydrogen peroxide is definitely not advised for use inside semi-closed cavities of the body, such as the bowel and the sinuses in your head. Because the oxygen gas that is released has nowhere to go it forces its way into the tissues, sometimes with devastating results.

And hydrogen peroxide is definitely not recommended at industrial-strength concentrations of 30% for use anywhere on, or in the body. It can burn the flesh at these concentrations.

The Cure? H_2O_2 (Inject It)

One of the most dangerous and incredibly risky ways to add H_2O_2 to your body is to inject it directly into your veins. In the bloodstream hydrogen peroxide has a half-life of 0.75–2 seconds before it turns into water, oxygen and various powerful radicals (very short-lived and very reactive chemicals).

In one case, a man with AIDS had read (but not in the medical literature) of the benefits of injecting hydrogen peroxide into his veins. The 35% hydrogen peroxide destroyed many of his red blood cells. The breakdown products of the red blood cells damaged his kidneys, and he died five days later with kidney failure.

Injecting hydrogen peroxide also creates a completely separate hazard — the formation of bubbles in the bloodstream. These bubbles take up to 30 minutes to dissolve into the liquid in the blood. However, the average circulation time of blood through the whole body is of the order of 10–15 seconds. These bubbles have been known to cause brain damage, by stopping the blood from getting to where it should be.

On 14 March 2004, a woman from South Carolina had hydrogen peroxide injected into her veins supposedly to treat her multiple sclerosis. She died.

Numbers – Oxygen Carried

Blood consists of salt water and red blood cells. The volume taken up by the red blood cells is about 35-47% in women and 40-54% in men.

The red blood cells carry the vast majority of the oxygen, while the salt water carries hardly any. In the average person, these cells are already carrying about 98% of the oxygen that they can potentially carry. You can see that there's not a lot of room for improvement – only another 2%.

By the way, you have to be pretty sick to have an oxygen level in the red blood cells of below about 92%.

The Cure?
H_2O_2 (Drink It)

Another dangerous delivery method is to *drink* H_2O_2. There's a basic misunderstanding involved here. Your gut is not very good at gas exchange — this is the job of the lungs. Only a little oxygen will get into your blood via the lining of your gut but, unfortunately, it does so in a dangerous way.

One report in the medical literature discussed six children who drank hydrogen peroxide — one of them died. The others suffered bouts of vomiting and severe burns to the oesophagus. Another report discusses a man who drank 30 ml of 35% hydrogen peroxide. He had a stroke.

Other complications that have been reported include respiratory arrest, gas bubbles in the blood vessels draining the gut, diffuse haemorrhages and oedema of the lining of the stomach, brain oedema and, of course, many fatalities.

The Cure?
H₂O₂ (Other Ways ...)

Some practitioners of Oxygen Healing Therapy irrigate wounds with it. This can cause bubbles of gas in the skin (surgical emphysema) — a very nasty condition.

Some practitioners even give enemas of hydrogen peroxide. One report discusses three cases of people who, after having a hydrogen peroxide enema, came down with acute ulcerative colitis. Others reported nasty side effects including rupture, necrosis and gangrene of the bowel.

The benefits of hydrogen peroxide in Oxygen Healing Therapy are zero (or close to it), while the risks are high.

The Cure? O₃

The history of oxygen therapy probably began in the mid-1800s in Germany, when it was claimed that ozone (O_3) could 'purify' blood. This was, of course, never proved.

However, ozone therapy has become probably the second most popular way of delivering extra oxygen. Like hydrogen peroxide, the delivery medium is a chemical (ozone O_3) that has one more oxygen atom than regular oxygen (O_2). Ozone is an unstable gas which, at standard room conditions, will break down to regular oxygen with a half-life of about 45 minutes.

The pure gas is sometimes passed over open wounds supposedly to improve healing and magically dissolve tumours. It doesn't.

Sometimes 20–50 ml of an ozone/oxygen gas mixture is injected into the muscle or under the skin. This is supposed to cure cancers, as well as allergic and inflammatory diseases. It doesn't.

Sometimes the ozone/oxygen gas is injected directly into the veins or even the arteries — a very risky procedure. It reacts with the water in the blood to make hydrogen peroxide which, in turn, breaks down to produce oxygen gas, water and a whole bunch of very reactive chemicals.

The ozone/oxygen gas is sometimes passed directly into the

rectum. The aim is to pass about 2 litres per minute into the gut, hoping that it will be absorbed via the large intestine.

There's another risky method called 'autohaemotherapy'. This involves removing the blood, exposing it to ozone and then replacing it.

Each of the above delivery methods has its own risks — and no proven benefits. The medical literature has many articles about the serious problems that have affected people who took ozone.

Interestingly, during World War I, medical doctors used ozone to treat injured soldiers including those suffering from 'trench foot' and the effects of poison gas. The results were mixed.

The Cure? Germanium Sesqui-oxide

Germanium sesqui-oxide is a compound that supposedly offers numerous health benefits from anti-ageing to increased natural immune response. Its chemical formula is $(GeCH_2CH_2COOH)_2O_3$.

It was first made in 1967 by a Japanese metallurgist, Kazuhiko Asai. The US Food and Drug Administration evaluated it and, in 1988, declared that it was not allowed to be used as either a food or a drug.

This did not stop GINA (Germanium Institute of North America) from publicising it. This institute seems to be run by a zoologist, a geneticist and a businessman who was previously involved in 'questionable cancer treatments' in Mexico. No medical people appear to be involved.

This mineral appears to have no proven benefits. On the other hand, it has serious, well-documented side effects such as kidney failure.

The Cure? Hyperbaric Chamber

A hyperbaric chamber is a strong container able to take one or more people, while the internal pressure is increased to several atmospheres.

In the 1940s, these chambers were used by the military to treat deep-sea divers with decompression sickness (the bends). In the 1950s, they were used for treating carbon monoxide poisoning. They are also used to treat delayed bone damage due to radiation (osteoradionecrosis). In this case, the chamber is filled with pure oxygen at 1.5–3 atmospheres of pressure. This therapy, known as HBOT (Hyperbaric Oxygen Therapy), is also used medically to treat arterial gas embolism — bubbles of gas in the arteries.

But the 'oxygen healing' uses of hyperbaric chambers are much wider. They are claimed (again with no proof) to treat chronic fatigue syndrome, allergies, AIDS, arthritis, autism, senility, gastric ulcers and just about any disease mentioned on the Internet. Each session lasts between 30 minutes and two hours, operating at a pressure of about 2.5 times atmospheric pressure.

Side effects of this therapy include fatigue, headaches, short-sightedness, damage to the sinuses in the head, lung damage and ruptured eardrums. Patients with certain lung diseases can suffer a collapsed lung. And, of course, the well-known complication called 'oxygen toxicity' can result in seizures, respiratory failure and fluid in the lungs. And don't forget the mechanical safety — fires or explosions in these hyperbaric chambers have killed over 80 people worldwide. You really need a good reason to choose this treatment, always ensuring that the benefits outweigh the risks.

About the only 'science' in favour of Hyperbaric Oxygen Therapy for Oxygen Healing Therapy dates back to 1900. It was then that Orville Cunningham noticed that some patients with heart disease seemed to do better if they lived near the ocean (where the air pressure is greatest) rather than in the mountains (where the air pressure is less). Yes folks, this is the entire 'science' behind hyperbaric oxygen healing.

The Cure? Oxygen Drinks

Oxygen drinks are ridiculously easy to summarise. They simply don't work. They are supposed to improve your health, stamina, sexual prowess or immune system, but all they do is empty your wallet.

Dr C.A. Piantodosi wrote that the rationale behind these drinks 'conflates physics and physiology in a struthonian visit to placebo land'. (Lost? Well let me explain it to you. The word 'conflate' means 'combine', while 'struthonian', from the Latin word *struthio* meaning 'ostrich', means 'ostrich-like' or 'ignorant of the facts'. In other words, the rationale combines two sciences in a sham kind of therapy that ignores the facts.)

First, he worked out the maths to show that a litre of hyperoxygenated water would carry less oxygen than a breath of fresh air. Want more oxygen? You have two choices — buy a bottle of bubbly water or take an extra breath. I know what I would choose.

Second, most of the stored oxygen in the bottle vanishes when you open the bottle and the bubbles explode to the surface.

Third, he quoted three separate studies that looked at the performance of athletes drinking regular or oxygenated water. The result? There was no difference.

The Cure? Oxygen Bars

Oxygen bars, like oxygen drinks, seem to come and go in popularity. Once again, they have no real benefits.

To make it more palatable for paying customers, the oxygen gas used in these bars is often given a flavour e.g. peppermint, wintergreen, cranberry or bayberry.

However, the oxygen gas can have some very dangerous side effects. Some people — including those with heart disease, asthma, congestive heart failure, pulmonary hypertension, hypertension and chronic obstructive pulmonary disease — can actually stop breathing if they inhale lots of pure oxygen. People who already have a cancer for which they are taking a drug called bleomycin can suffer lung damage. And if the flavouring of the oxygen gas is oil-based, it can cause lipoid pneumonia in some people.

The flavoured oxygen in oxygen bars is definitely not a pure, medical-grade oxygen, such as Aviatiors Breathing Oxygen (ABO).

Lies, Lies and More Lies

All drugs have side effects. So if you intend to use a drug, you need to be sure that the benefits outweigh the risks. This is definitely not the case with Oxygen Healing Therapy.

The bad science held up as proof of oxygen healing treatments can be seen in one so-called 'study' from the Dominican Republic which claimed to have used ozone to cure 13 people of cancer.

During a follow-up, two people refused to be interviewed, three more could not be proved to have ever had cancer, and three could not be proved to have ever existed. Of the remaining five, three were alive but with cancer and two had already died of cancer.

A cure ratio of 0% is pretty appalling. However, if you're hoping for better results from oxygen therapy, don't hold your breath ...

References

American Cancer Society, 'Questionable methods of cancer management: hydrogen peroxide and other "hyperoxygenation" therapies', *CA: A Cancer Journal for Clinicians*, January/February 1993, pp 47–56.

Eccles, Kathryn, 'Oxygenated water theory full of hot air', *Medical Observer*, 8 September 2006, p 9.

Green, Saul, 'Oxygenation therapy: unproven treatments for cancer and AIDS', *The Scientific Review of Alternative Medicine*, Spring/Summer, 1998, Vol 2, No 1, and also on http://www.quackwatch.org/01QuackeryRelatedTopics/Cancer/oxygen.html.

Hirschtick, Robert E., et al., 'Death from an unconventional therapy for AIDS', *Annals of Internal Medicine*, 15 April 1994, p 694.

Meyer, Christopher T., et al., 'Hydrogen peroxide colitis: a report of three patients', *Journal of Clinical Gastroenterology*, 1981, Vol 3, pp 31–35.

Piantodose, C.A., '"Oxygenated" water and athletic performance', *British Journal of Sports Medicine*, 2006, Vol 40, p 740.

Sherman, S.J., et al., 'Cerebral infarction immediately after ingestion of hydrogen peroxide solution', *Stroke*, May 1994, pp 1065–1067.

Galileo and the Leaning Tower of Pisa

From the beginning of time people believed that heavy objects fall faster than lighter objects. After all, Aristotle said so. The story goes that to challenge this belief, a few centuries ago Galileo climbed to the top of the Leaning Tower of Pisa and dropped a large, heavy cannon ball and a small, light musket ball over the edge at the same moment. Both balls hit the ground at the same time and, with this elegant experiment, Galileo resoundingly debunked Aristotle's theory. Everybody has heard this fantastic story. It's simple and straightforward, but it's not correct.

History of Falling Bodies

The story begins more than 2000 years ago with Aristotle (384–332 BC). One of the greatest intellects ever produced by Greece, he mastered most of the known sciences and arts of his time.

He claimed that stones fell to the ground because they were made mostly of 'earth' and were therefore naturally attracted to the earth. On the other hand, he claimed that because smoke was made mostly of 'fire' and 'air' it tried to rise to its natural home in the heavens. This is not correct.

This will make your balls drop

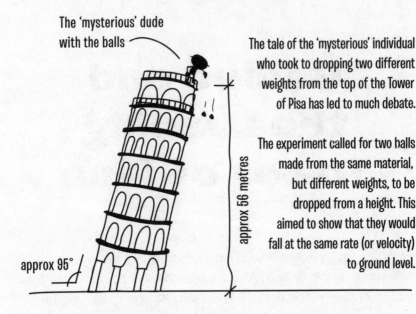

The 'mysterious' dude with the balls

The tale of the 'mysterious' individual who took to dropping two different weights from the top of the Tower of Pisa has led to much debate.

The experiment called for two balls made from the same material, but different weights, to be dropped from a height. This aimed to show that they would fall at the same rate (or velocity) to ground level.

approx 56 metres

approx 95°

The astonishing Leaning Tower of Pisa

However, Aristotle was 100% correct when he stated that a body would fall more slowly in a thick medium (water) than in a thin medium (air). He also correctly said that in a very thick medium (oil) a heavy body would fall faster than a light body. And again, he correctly said that in a vacuum bodies of all sizes would fall at the same speed.

But then he blotted his copybook by claiming that a vacuum was impossible with his famous words, 'Nature abhors a vacuum'. And unfortunately, he wrongly summarised his own work in Falling Bodies Theory by saying that the speed of a falling body would depend on its bulk.

About 500 years after the birth of Christ a scientist from Alexandria called John Philoponus (a.k.a. John the Grammarian and John of Alexandria, 490–570 AD) set the record straight. He dropped two very different weights from a tower and found that they landed at almost the same time.

Leaning Tower of Pisa

The Leaning Tower of Pisa comes into the story in 1173 AD, the year when construction began. By the time the tower reached the third storey it had begun to lean, thanks to the soft ground that it was built on. The engineer and architect Bonnano Pisano tried to correct for this by making the remaining storeys shorter on the uphill side. But this didn't help. The Leaning Tower of Pisa kept on leaning more and more and by the time it was closed for repairs in 1990, it was leaning 5.5° from the vertical, the top floor overhanging the base by 4.5 m. In fact, by this time it was in real danger of collapsing. The modern-day repairs were finished in 2001, reducing the overhang to 4.1 m, so the Leaning Tower of Pisa should hang in there for a few more centuries.

And what about our hero, Galileo (1564–1642)? Galileo was born in Pisa, matriculated in 1581 to study Medicine at the University of Pisa and took up the Chair of Mathematics there in 1589. Although he would have had access to the tower he almost certainly did not drop two balls from its 56-m high top floor. There is absolutely no believable, contemporary record of his having done so.

Who Dun It?

Galileo himself did not ever say that he had dropped the two different weights from the Leaning Tower of Pisa.

Robert P. Crease — from the Department of Philosophy at the State University of New York at Stony Brook, and historian at the Brookhaven National Laboratory — has a theory about this claim. He blames Vincenzio Viviani (1622–1703), Galileo's secretary during his final years.

Viviani wrote an affectionate biography of Galileo in which he described how Galileo climbed the Leaning Tower of Pisa and carried out the famous experiment 'in the presence of other teachers and philosophers and all the students'. And, having done the experiment many times, Galileo was able to prove to the others that 'the velocity of moving bodies of the same composition, but of different weights, moving through the same medium, do not attain the proportion of their weight as Aristotle decreed, but move with the same velocity'. Viviani is the first writer to claim that Galileo dropped the two balls from the Leaning Tower of Pisa. Galileo did not ever make such a claim — and there are *no other* contemporary reports of this event.

It is a lovely experiment. In fact, experiments involving falling bodies made it into the Top 10 in a survey by Robert Crease of the 'most beautiful' experiments.

Because dropping balls from the Leaning Tower of Pisa was such a lovely concept, it then gained a life of its own.

Science historian Paolo Palmieri, of the University of Pittsburg, has another theory. He suggests that only very late in life did Galileo arrive at his conclusion that all bodies fall at the same rate in a vacuum. Palmieri says that perhaps his friends were a little too affectionate and tidied up his long and messy experiments into a simple experiment.

Galileo's Way

Galileo approached the problem of freely falling bodies in an indirect way, beginning with thought experiments, i.e. hypothetical experiments that use 'what if' type of reasoning.

In the early 1590s, he claimed that bodies with the same density would fall at the same speed. This is probably correct, depending on the conditions.

Soon afterwards — following more thought experiments — he concluded that bodies of different densities would fall at different speeds in a vacuum. Unfortunately, on this one he was very wrong. He then moved on from thought experiments to actual experiments.

He spent the next half century rolling balls down ramps and developing theories of fluid dynamics. And finally, after looking back over all of his experimental data, he correctly concluded that bodies of any density would fall at the same speed in a vacuum.

Thought Experiment Contradiction

Thought experiments can lead to interesting conclusions. Let's see what happens with Aristotle's idea that light bodies fall slowly and that heavy bodies fall quickly.

Suppose you have a 4-kg weight and an 8-kg weight. According to Aristotle, the 4-kg weight will fall more slowly than the 8-kg weight. The 4-kg weight 'naturally' falls more slowly.

Now tie the two weights together, making a 12-kg weight.

Looking at it one way, the 4-kg weight will 'naturally' fall more slowly, thus slowing down the 8-kg weight. Therefore, the two weights combined into one (i.e. 12 kg) will fall more slowly than the 8-kg weight.

Looking at it another way, the combined weight is 12 kg. It is heavier than either the 4-kg weight or the 8-kg weight. Therefore, the 12-kg weight will fall more quickly than the 8-kg weight.

What a lovely contradiction! Maybe our starting assumptions are wrong?

Galileo put forward this very paradox in his book of 1638, *Two New Sciences*, in a discussion between Salviati and Simplicio.

The Final Proof

On 2 August 1971, the *Apollo 15* astronaut David Scott performed the actual experiment on the Moon. He dropped a geology hammer and a falcon feather at the same instant. (He chose this feather because his spacecraft was called the *Falcon*. Who says that scientists are not poetic souls?) In the deep vacuum of Space, both objects hit the lunar surface at the same time.

Aristotle would be tickled pink to think that it would take 2300 years for the hammer to fall on his theory.

The Exact Words from the Moon

Commander David R. Scott (2 August 1971, lunar surface): 'Well, in my left hand I have a feather; in my right hand, a hammer. And I guess one of the reasons we got here today was because of a gentleman named Galileo, a long time ago, who made a rather significant discovery about falling objects in gravity fields. And we thought: "Where would be a better place to confirm his findings than on the Moon?".'

[Camera zooms in on Scott's hands. One is holding a feather, the other a hammer. The camera zooms back to show the *Falcon*, the *Apollo 15* landing craft – and the lunar horizon.]

Scott: 'And so we thought we'd try it here for you. The feather happens to be, appropriately, a falcon feather for our *Falcon*. And I'll drop the two of them here and, hopefully, they'll hit the ground at the same time.'

[Scott releases the hammer and the feather which hit the ground at about the same time.]

Scott: 'How about that! Mr Galileo was correct in his findings.'

References

Atkinson, David, et al., 'Galileo and prior philosophy', *Studies in the History and Philosophy of Science*, 2004, Vol 35, pp 115–136.

Crease, Robert P., 'The legend of the Leaning Tower', physics website associated with *Physics World*, February 2003, http://physicsweb.org/articles/world/16/2/2/1.

Galilei, Galileo, *Discorsi e dimostrazioni matematiche, intorno a due nuove scienze (Dialogues concerning two new sciences)*, translated by H. Crew and A. de Salvio, New York: Dover Publications, 1954 (first published 1638), p 108.

Palmieri, Paolo, '"*Spuntar lo scoglio piu duro*": did Galileo ever think the most beautiful thought experiment in the history of science', *Studies in the History and Philosophy of Science*, 2005, Vol 36, pp 223–240.

Can Diamond Cut It?

In 1948, a copywriter for the N.W. Ayer advertising agency came up with the slogan 'Diamonds Are Forever'. In 2000, this slogan was judged by the trade journal *Advertising Age* to be the best slogan of the 20th century. Diamonds, which are really just hard white shiny stones, have a special mystique thanks to some very clever and long-running advertising. So it's inevitable that some of the stories about diamonds will be wrong.

One myth suggests that diamond is the only gem that can scratch glass. The flip side of this myth is that 'fake' diamonds cannot scratch glass.

Atomic Structure of Diamonds

For years, only two forms of pure carbon were known — graphite (the soft, black stuff in pencils) and diamond (the hard, transparent stuff in engagement rings).

In graphite, the carbon atoms are arranged into flat sheets, stacked on top of each other. There are many chemical bonds inside each sheet, but only very few joining one sheet to the next. This means that the sheets can slide easily over each other, making graphite a good lubricant. In fact, graphite will lubricate the barrel of a sticky lock much better than oil. Oil will attract dirt that will ultimately clog the lock. But graphite won't.

Diamond is also made of carbon atoms that are packed into a very tight, cubic crystal structure — with tiny pyramids, each packed with only five carbon atoms, stacked on top of each other. If you apply a load anywhere on the surface of a diamond, this pyramid structure spreads the load through the whole body of the diamond making it very hard.

In 1985, we discovered that carbon atoms can arrange themselves into yet another shape. Called a buckyball or fullerene, it has 60 carbon atoms arranged into the shape of a microscopic soccer ball 0.7 nm (a billionth of a metre) in diameter.

History of Diamonds

The Ancients were fascinated by diamonds. They are mentioned in various Buddhist and Sanskrit texts as far back as 300 BC. In India, they were regarded as religious icons. Indeed, India led the world in diamond production from the 9th century BC to the mid-18th century. Diamonds were found only in hollows in river beds in India and, much later, in Brazil. Back then, the annual diamond production for the whole world was only a kilogram or two.

These strange gems were the hardest things found anywhere in the world and virtually indestructible. Because this shiny gem was so tough, the Greeks called it *adamas* meaning 'unconquerable' or 'invincible' from which we get the words 'diamond', and 'adamant'.

Grinding Wheels

The toughest grinding wheel possible is made out of diamond — not transparent, clear gems which are too expensive to use, but coloured, industrial diamonds which are much cheaper. Their colours come from impurities, like nickel or cobalt, that have infiltrated in between the carbon atoms.

A diamond grinding wheel can cut through almost anything in the known Universe.

Buckyballs

The buckyball (or fullerene) gets its names from the architect Richard Buckminster Fuller, who around 1950 devised and popularised geodesic domes. It follows the Polyhedron Formula put forward by brilliant mathematician, Leonhard Paul Euler (1707–1783). This formula covers all shapes that are made up of flat surfaces with straight edges, and that are roughly spherical – and that don't fold in on themselves. It covers most shapes from a pyramid up. The formula is:

$$V - E + F = 2$$

where   V = number of vertices

E = number of edges

and   F = number of faces.

Just like a soccer ball, a buckyball consists of 12 pentagons and 10 hexagons.

Leonhard Euler published more scientific papers than any other mathematician in recorded history. He worked in geometry, trigonometry, algebra, number theory, calculus, continuum physics, lunar physics, and so on. He began to go blind in his early thirties. He compensated for this with his amazing memory.

Since the first 'soccer ball' fullerene was discovered in 1985, many others have been discovered. There are many spherical fullerenes (with both more and fewer carbon atoms), and fullerenes with different shapes, such as hollow rods. These hollow rods are incredibly strong, and in some cases, harder than diamond. They are only billionths of a metre across, but up to several millimetres long.

Free Diamonds on Venus

If you would like two superb, large diamonds, each the size of a 10-cent piece, why not go to Venus? You don't have to dig underground to get them – they are nicely giftwrapped on two decaying Soviet spacecraft.

The Soviets are the only ones to ever successfully land a robotic spacecraft on the surface of Venus – a truly hostile environment. The temperature is up around 500°C, the atmospheric pressure is around 1000 tonnes per square metre (100 times greater than that on Earth), while the atmosphere itself is made of carbon dioxide and sulphuric acid.

This environment will quickly damage any spacecraft, including the front lens of any camera. And the Soviets wanted to be able to take photographs for the half-hour or so that the spacecraft would survive after it landed. For this reason the Soviets used a large block of top-grade, gem-quality diamonds as a window to protect the lens. These diamonds will survive the environment of Venus indefinitely.

So if you want a free diamond, you know where to get one – from Russia with love.

Diamonds from Space

Scientists have found diamonds inside meteors that have crash-landed on Earth. The meteors are older than our planet, which means that the diamonds are even older again. They are very small, about one-fifth of a micron (a micron is a millionth of a metre) in diameter — about 70 times smaller than the thickness of a human hair. Just the right size for a mosquito's engagement ring perhaps?

But where did these ancient mini-sparklers come from? From those giant hydrogen bombs in the sky — stars.

Bling, Baby, Bling!

Nature makes diamonds around 200 kilometres below the Earth's surface where the temperature is high enough (approx 900–1300°C) and where the pressure is around 45 000–60 000 atmospheres.

Where diamonds form

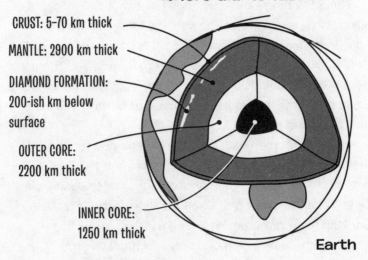

CRUST: 5–70 km thick

MANTLE: 2900 km thick

DIAMOND FORMATION: 200-ish km below surface

OUTER CORE: 2200 km thick

INNER CORE: 1250 km thick

Earth

How kimberlites form

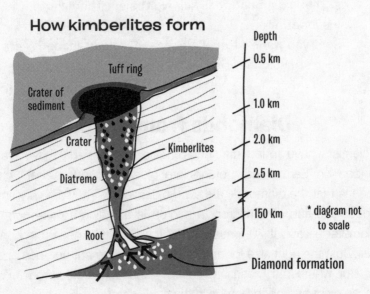

Tuff ring

Crater of sediment

Crater

Diatreme

Kimberlites

Root

Depth

0.5 km

1.0 km

2.0 km

2.5 km

150 km

* diagram not to scale

Diamond formation

Astrophysicists believe that these tiny diamonds are made in the thin atmospheres of stars, as the carbon atoms smash into each other at huge speeds. If this sounds unbelievable, just remember that just about every atom of gold in the Universe was made inside an exploding star, when lighter elements got smashed into each other.

Making Diamonds

In one of the Superman movies, our Man of Steel squeezes some coal in his super fist, turning it into a huge sparkler. Yup, temperature and pressure will turn carbon into diamond — temperatures above 400°C and 30 000 times our atmospheric pressure. The exact temperature and pressure depend on the process used.

In the 1920s, John Logie Baird (one of the inventors of television) tried to use temperature and pressure to make diamonds. He filled a 44-gallon drum (about 205 litres) with dynamite, concrete and coal — the dynamite to provide temperature and pressure, the concrete to trap the temperature and pressure, and the coal to turn into diamonds. Yes, he did explode it. But no, he did not find any diamonds scattered over the blast area of about 100 m^2.

It was not until the mid-1950s that the General Electric Company in New York succeeded in making the first useable synthetic diamonds. They began making tiny industrial diamonds, using temperatures of 1700°C and pressures of 70 000 atmospheres. We can now make diamonds out of virtually any source of carbon, including methane.

Industrial diamonds are used for machine tools, dies to draw wires, mining drills, quarrying saws and grinding wheels. Today, it's relatively easy to make tiny gems but they are still more expensive to make than to dig the natural ones out of the ground.

Natural Diamonds

Nature makes diamonds about 200 km underground, where the temperature and pressure are high enough. At these depths the temperature is about 900–1300°C and the pressure around 45 000–60 000 atmospheres.

But how do diamonds get to the surface?

At 200+ km below ground, the temperature and pressure are high enough for a diamond to stay a diamond. And at the surface, the temperature and pressure are low enough for a diamond to be stable.

However, in between, where the temperature is still very high but the pressure is not high enough, a diamond will turn into graphite. That's fine if you want to make a 'lead' pencil, but not if you want diamonds for a drilling tip.

Physical chemists tell us that the only way for diamonds to survive the trip to the surface is to do it quickly — in no more than a few hours. But how can diamonds possibly move through rock, even if it's liquid, at speeds of 100–200 kph?

A clue comes from the Russians, and now the Iranians, who both claim to have torpedoes that can move through water at 400 kph. This torpedo apparently blows a gas out in front of it, so that the metal body of the torpedo never makes contact with the water — it touches only the gas. In turn, the gas pushes through the water at 400 kph, the torpedo following closely behind.

Kimberlite Pipe = Fabulous Wealth

The new torpedo technology could be the principle behind the strange geological phenomenon called a kimberlite pipe, named after the town of Kimberly in South Africa, where the first such pipe was discovered. Shaped like a funnel, it's typically 100–1000 m in diameter at the top with a very long stem that goes down through the solid crust of the Earth to the top of the liquid rock of the semisolid mantle.

Although there are lots of kimberlite pipes they are hard to find, and only one in every 200 has lots of diamonds. It is then simply a matter of digging into the kimberlite ore, bringing all the rock to the surface and extracting the diamonds. On average, two tonnes of blueish kimberlite ore will produce one carat of diamonds. The quantity of diamonds decreases with increasing depth, because the ore body tapers off as it gets deeper.

In Kimberley in South Africa, the Big Hole was dug out to a depth of just over 1 km before it was closed in 1914. It is still the largest excavation ever dug by hand.

Carat = Carob

The weight of diamonds is measured in carats and points.

A carat is 200 mg, i.e. one-fifth of a gram (defined in 1907 by the Fourth General Conference on Weights and Measures). A point is one-hundredth of a carat.

This measure apparently comes from the ancient practice of measuring the weight of diamonds against carob seeds. It was believed that all carob seeds were identical, supposedly making them perfect as 'standard' weights.

The carob tree is an evergreen Mediterranean tree that bears long, brown-purple pods that look like snow-pea pods. Just like a snow pea, there are little seeds inside. (I first ran across carob pods in health food shops, as a 'healthy' substitute for chocolate. It is also called St John's Bread, because John the Baptist supposedly kept himself alive by eating carob pods in the desert.)

In the diamond trade, carob seeds were considered to be the same weight, regardless of the size of the pod, the size and age of the tree, and the location and growing conditions of the tree. However, a study in 2006 found that there is actually a huge variation in the weights of carob seeds. Amazingly, the average weight of a carob seed was 200.5 mg, but individually they varied between 140 and 230 mg. Perhaps the ancient diamond traders simply chose the carob seed because it was widely available around the Mediterranean.

More on the formation of bling ...

Kimberlite Pipe = High-Speed Elevator

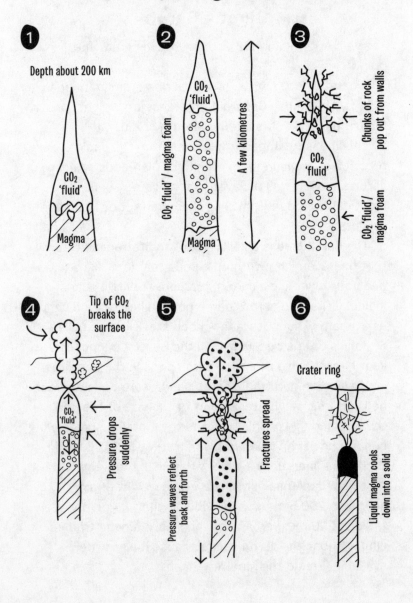

1 Depth about 200 km

CO$_2$ 'fluid'

Magma

2 CO$_2$ 'fluid' / magma foam

CO$_2$ 'fluid'

Magma

A few kilometres

3 Chunks of rock pop out from walls

CO$_2$ 'fluid'

CO$_2$ 'fluid'/ magma foam

4 Tip of CO$_2$ breaks the surface

CO$_2$ 'fluid'

Pressure drops suddenly

5 Pressure waves reflect back and forth

Fractures spread

6 Crater ring

Liquid magma cools down into a solid

Kimberlite Pipe = High Speed Elevator

So how do the diamonds get to the surface?

One theory suggests that at 200+ km below ground a rocket-shaped mass of supercritical carbon dioxide (CO_2) fluid forms in a temporary opening in the mantle. ('Supercritical' means that the carbon dioxide has properties of both a gas and a liquid.) Underneath the fluid carbon dioxide is a low-density mass of molten rock, magma. It's buoyant because it's riddled with carbon dioxide.

The buoyant mass of magma starts heading for the surface. As it gets closer, the pressure drops to a level such that carbon dioxide is released from the magma flooding into the space between it and the CO_2 fluid, creating a 'foam' of carbon dioxide and magma. This rocket-shaped mass of carbon dioxide/magma foam and CO_2 fluid is several kilometres long, and keeps moving upwards. As it rises, the tip of the CO_2 fluid fractures the rock. Chunks of rock fall from the walls through the CO_2 fluid and the foam mixture into the magma. The upward speed increases up to 180 kph.

As the tip of the fluid breaks the surface it evaporates due to the sudden drop in pressure. A blast of CO_2 gas, magma and chunks of rock then spurts up at speeds of up to 4000 kph.

When the surface hole closes due to the sudden cooling, a pressure wave spreads downwards and then bounces back up again. The rock near the surface is fractured, another blast then erupting into the atmosphere. An hour later, it's all over. A crater is formed at the surface and lots of diamonds are safely buried in a funnel-shaped lump of kimberlite, a few kilometres deep.

Tonnes of Diamonds

Worldwide, the current annual diamond production is more than 120 tonnes per year. Mines produce about 20 tonnes of industrial diamonds and 6 tonnes of gems, while laboratories make about 100 tonnes of synthetic diamonds.

Hardness

Is it true that fake diamonds are unable to scratch glass? To find out you need to consult a Table of Hardness.

Although there are a few different scales that measure the 'hardness' of minerals, the Mohs Scale — devised in 1812 by the German mineralogist Friedrich Mohs — is as good as any. It has 10 levels, ranging from 1 (talc, which is very soft) to 10 (diamond, which is the hardest).

Of course, some diamonds are 'harder' than other diamonds. Some of the hardest diamonds in the world come from the New England region in New South Wales in Australia.

And yes, diamond will scratch glass but so will other natural gems. For example, crystal quartz (Mohs Hardness of 7) will also scratch glass easily. Topaz (Mohs Hardness of 8) will scratch very easily, while corundum (i.e. rubies and sapphires, Mohs Hardness of 9) will actually cut glass.

Cubic zirconia, a very common 'diamond simulant' (a fancy name for a 'fake diamond'), is made not from carbon but from zirconium oxide. With a Mohs Hardness of 8.5–9, it can very easily scratch and sometimes cut glass. Moissanite (silicon carbide), another common diamond simulant, has a Mohs Hardness of 8.5–9.25. These two synthetic gems that look like diamonds can very easily scratch glass.

The Mohs Scale

'Hardness' is the resistance of a material to being permanently deformed, such as being scratched or abraded. Friedrich Mohs, the German mineralogist, devised the Mohs Scale – a 10-point scale with '1' the softest material and '10' the hardest.

1	Talc – easily scratched by a fingernail
2	Gypsum – just scratched by a fingernail
2.2	Fingernail

2.5–3	Pure gold and silver
3	Calcite – very easily scratched by a metal knife, and just scratched by a copper coin;
3.2–3.5	Copper coin
4	Fluorite – not hard enough to scratch glass, but hard enough to scratch a copper coin
4–4.5	Platinum
5	Apatite – just hard enough to scratch glass and can be scratched by a steel knife with difficulty
5.1	Pocket knife
5.5	Glass
6	Orthoclase feldspar – easily scratches glass and can be just scratched by a metal file
7	Quartz (silicon oxide, tiger's-eye, amethyst, citrine, aventurine) – too hard to be scratched by a metal file but scratches glass easily
8	Topaz – scratched only by corundum and diamond, but scratches glass very easily
8.5–9	Cubic zirconia (zirconium oxide)
9	Corundum (aluminium oxide, sapphires, rubies) and moissanite (silicon carbide) – scratched only by diamond
10	Diamond – scratched only by another diamond and ultra-hard fullerite and aggregated diamond nanorods
(10 + a bit)	– synthetic materials such as ultra-hard fullerite and aggregated diamond nanorods

The Mohs Scale is not linear. If it were linear, and if corundum were still rated at 9, then diamond would be rated at 42. And, if calcite was still 3, then fluorite would be 3.25, not 4.

Another complication is that diamond appears to vary in hardness, depending on the angle that the deforming force is applied.

Harder Than Diamond?

Over the past decade, science has busted another lovely myth. Diamond is no longer the hardest substance.

In 1997, ultra-hard fullerite (buckyballs of C_{60}) was made in a laboratory. At that moment, diamond lost its title as 'Hardest Stuff in the Universe'. One way to measure 'hardness' is by the pressure (measured in gigapascals) needed to permanently mark or deform it. Diamond clocked in at about 170 GPa, while the ultra-hard fullerite was about 300 GPa.

In 2005, German researchers subjected C_{60} fullerite balls to 200 atmospheres and 2252°C. In this process, they manufactured another form of carbon – aggregated diamond nanorods. Each rod is about 5–20 nm (billionths of a metre) in diameter, and about 1 micron long. These rods could withstand about 490 GPa.

How to Test a Diamond

Diamond has one of the highest thermal conductivities of any material known, i.e. it can conduct heat faster than virtually any other substance known.

This gives you an easy way of testing if an unknown gem is a diamond. Breathe on it, to give it a surface fog. Because a lump of glass or a cubic zirconia conducts heat fairly slowly, it will hang onto the fog for a while. A real diamond, however, will get rid of the surface fog almost immediately. Unfortunately, so will the diamond simulant known as moissanite.

The best way is to take the gem to an independent jewellery appraiser.

But whatever you do, don't trust the scratch-glass test.

Is a Diamond Forever?

Do diamonds really last forever?

As long as you don't try to shatter, chip or burn them, they do. This is why the famous slogan was invented — to reduce the 'aftermarket'. If people could be convinced to hang onto their old diamonds and not try to sell them, then diamond retailers could keep on selling freshly mined diamonds. In fact, before the famous advertising campaign, exotic coloured gems such as topaz, rubies and sapphires were the gems of choice in many engagement rings — not diamonds.

So diamonds can last forever — and being incredibly hard has to help. And, it's still true that 'diamonds are a girl's best friend' …

Largest Diamond

The largest diamond ever discovered was the 3106.75-carat Cullinan diamond found in 1905 in Transvaal in South Africa. However, because it had a few flaws it was cut into nine large stones. One, the 530.2-carat Star of Africa diamond, is set in the English royal sceptre.

Advertising Slogans Do Work!

The trade publication *Advertising Age* ran a survey to find the Top 10 advertising slogans of the 20th century. The top three were:

1 Diamonds are forever (DeBeers)
2 Just do it (Nike)
3 The pause that refreshes (Coca-Cola)

References

Blank, V., et al., 'Nano-sclerometry measurements of superhard materials and diamond hardness using scanning force microscope with the ultrahard fullerite C60 tip', *Journal of Materials Research*, November 1997, pp 3109–3133.

Blank, V., et al., 'Ultrahard and superhard phases of fullerite C_{60}: comparison with diamond on hardness and wear', *Diamond and Related Materials*, 1998, Vol 7, pp 427–431.

Del Re, Gerard, *The Whole Truth: A Compendium of Myths, Mistakes and Misconceptions*, New York: Random House, 2004, p 198.

'Did carob seeds allow shady diamond deals?', *New Scientist*, 9 May 2006, p 20.

Dubrovinskaia, N., et al., 'Aggregated diamond nanorods, the densest and least compressible form of carbon', *Applied Physics Letters*, 2005, Vol 87, pp 083106–1 to 083106–3.

Epstein, Edward Jay, *The Diamond Invention*, Arrow (A Division of the Random House Group), 1982.

Perkins, Sid, 'A gemstone's wild ride: how diamonds erupt from deep within earth', *Science News*, 30 June 2007, p 412.

Turnbull, L.A., 'Seed size variability: from carob to carats', *Biology Letters*, 2006, Vol 2, pp 397–400.

Wilson, L., et al., 'An intergrated model of kimberlite ascent and eruption', *Nature*, 3 May 2007, pp 53–57.

Younger-Dickinson, Joan, *The Book of Diamonds, Their History and Romance from Ancient India to Modern Times*, New York: Avenel Books, 1965, pp 36, 41, 50, 51.

Reincarnation - Again

As part of my search for knowledge, I read newspapers. So my interest was aroused when I read in the *Los Angeles Times* that after six-and-a-half years of 'scientific research', the one, true and you-betta-believe-it reincarnation of Marilyn Monroe had been discovered. However, this reincarnation is by no means the first.

Reincarnation 101

The word 'reincarnation' literally means 'to be made flesh again', while the dictionary definition refers to the 'rebirth of a soul in a new body'. In other words, some part of a person survives beyond their death to be reborn in a brand new body.

Although reincarnation is a tenet in some Indian, African, Eastern and Native American religions, it was never a major cornerstone of traditional Western religions or mainstream Islam. However, a few ancient Greek philosophers believed in reincarnation. Both Pythagoras and Plato could recount details of their 'previous' lives, while Socrates said in his later years: 'I am confident that there truly is such a thing as living again, and that the living spring from the dead.'

But, there is no hard scientific evidence for reincarnation — yet.

There are a few problems with the concept of reincarnation, as we know it in Western society.

First, why is it mainly famous people — Cleopatra, Napoleon, Marilyn Monroe — who are reincarnated? Why don't ordinary people get reincarnated?

Second, how can you have more than one reincarnation of the same person at the same time? At any given time, there seem to be hundreds of people each claiming to be Marilyn, Cleopatra or Napoleon.

Third, if we are all reincarnated, and if the population of our planet is growing, where do the extra souls come from? Very good insects?

Reincarnation and Health

Somehow, the reincarnation industry has become married to the health industry.

There are some reincarnation tests that use your physical symptoms to work out what happened to you in a previous life. For example, they claim that persistent neck problems point to the fact that you were probably hanged in a previous life, while a fear of heights means that you were once thrown off a castle wall in the 15th century.

Latest Marilyn Monroe

This latest Marilyn Monroe reincarnation is Sherrie Lea Laird, the lead singer of the Canadian rock band Pandamonia. Just in case you have never heard of her, she is modestly claimed to be at the top of the recording charts in Canada, Europe and, in fact, the Whole World. However, I have never been able to find her in iTunes.

Her claim to being Marilyn Monroe's reincarnation was researched (if this is the correct word) by the Malibu psychiatrist, Dr Adrian Finkelstein. He chose to first publish his work, not in the

New England Journal of Medicine or *The Lancet*, but in the 26 May 2005 issue of *Malibu Surfside News*.

His 'scientific research' consists of many hours of video tapes of Ms Laird (while she was hypnotised) and noting similarities between Ms Laird and Ms Monroe, e.g. facial bone structure and 'handwriting and personality traits assets and liabilities'. Dr Finkelstein also claims that Ms Laird was able to provide accurate information about Marilyn, information that was known only by Marilyn. If Marilyn was the only person who knew this information, how did he, Dr Finkelstein, also know whether this information was right or wrong? Perhaps the many hours of phone conversations that Ms Laird had with Ted Jordan, a close associate of Marilyn Monroe, may have expanded her knowledge of Marilyn's private life.

Many desire to be made flesh again

'Reincarnation' literally means 'to be made flesh again'. The dictionary defines it as the 'rebirth of a soul in a new body'. In other words, some part of a person survives beyond their death to be reborn in a brand new body.

Obscure Tests for Past Lives

On his home page, Adrian Finklestein modestly refers to himself as 'pioneer and world foremost expert in Past Life Regression Therapy'.

To assist him in this ground-breaking work, he has invented some new tests. They include PPL-IRCT (Past and Present Lives Iris Recognition Comparison Test), PPL-VRCT (Past and Present Lives Voice Recognition Comparison Test) and, of course, PPL-FPRCT (Past and Present Lives Finger Print Recognition Comparison Test).

Finkelstein claims that his research shows that 'Sherrie Lee Laird is the linear, thus only reincarnation of Marilyn Monroe'.

Backing the doctor up was some very heavy recommendation from the Other Side. His home page (www.pastlives.com) claims that '... the ancient Egyptian spirit, Ahtun-Re, notorious for its accuracy, gave the same confirmation'. (To be frank, until I read this, I didn't have any concept of the accuracy of any ancient or modern Egyptian spirits.)

Reincarnation 102

Now let me take you back to the famous Bridey Murphy case (that's Bridey with an 'ey'), which really triggered Western interest in reincarnation.

It all began in 1952 in the town of Pueblo, Colorado, when a resident called Virginia Tighe (while under hypnosis) began talking with a thick Irish brogue, telling stories of her previous life as Bridey Murphy, who was apparently born in Ireland in 1798. Under hypnosis, Virginia spoke knowledgeably of various events in her previous incarnation as Bridey.

She had a mother called 'Kathleen', a friend called 'Kevin' and married a man called 'Sean Joseph Brian McCarthy'. She (Bridey, this is) was strongly chastised for scratching paint off a metal bed frame and, unfortunately, had died by falling down a flight of stairs. These events were supposedly supported by reporters who checked the records in Ireland. In 1956, Morey Bernstein, the person who hypnotised her, wrote the book *The Search for Bridey Murphy*, becoming fabulously wealthy from the sales of the book, records of the trance sessions and the movie rights.

Only later did it emerge that the 'reporters' who had 'investigated' the facts had not bothered to actually check, either in Ireland or in Virginia Tighe's childhood town. Had they done so, they would have discovered that, as a teenager, Virginia had lived across the road from a woman who had actually been born in Ireland — and her name was Bridie Murphy (that's Bridie with an 'ie')! She wouldn't have been hard to find. She was still living in the same house that she had lived in when Virginia lived across the road.

All of Virginia's stories of 18th century Bridey had come from the intermingled life experiences of 20th century Bridie, the Irish neighbour, and Virginia's own childhood.

Bridie (the woman from across the road) had a mother called Katherine (fairly close to Bridey's mum, 'Kathleen'), and a real son called Kevin.

'Brian' was the middle name of Virginia's real life husband, while 'McCarthy' was the name of a character that Virginia had played on stage at high school. Virginia had a sister who had fallen down the stairs (but had not died). In a separate incident, Virginia's brother had died when she was five. And Virginia had been soundly whipped for scratching the fresh paint from her metal bed frame.

Marilyn Gets Weird

Getting back to the latest reincarnation of Marilyn Monroe, there's a bizarre twist. Ms Top of the Pops, Sherrie Lea Laird (who has had various drug and psychiatric problems), now claims that Kezia, her biological daughter, is her mother (okay, the reincarnation of Marilyn's mother, Gladys Baker).

Her psychiatrist Adrian Finkelstein has now written a book, *Marilyn Monroe Returns: The Healing of a Soul?* And reincarnation might be all the rage again.

T. Lopsang Rampa

In 1956, a strange and exciting book became a bestseller. *The Third Eye* was the autobiography of a Tibetan Buddhist monk. It told how in his childhood, he was sent to a Tibetan monastery where the monks gouged a hole in his head and filled it with wood. He studied and then mastered Tibetan Buddhism and could, as part of his powers, levitate. He escaped the Chinese invasion of Tibet ending up in England.

However, genuine scholars of Tibetan Buddhism thought that he was a fraud. And he was! He turned out to be a plumber's mate, Cyril Henry Hoskins, who had never been to Tibet and who didn't have a hole in his forehead – but he was interested in the occult. To show that he was serious, he grew a beard, shaved his head and called himself Dr Kuan-suo. Eventually he changed his name to T. Lopsang Tampa – the 'T' standing for Tuesday, because Cyril had been born on a Tuesday.

But Cyril could explain it all. Sure, he might have been born Cyril, but Rampa's spirit had chosen to reincarnate itself in Cyril's body – therefore, he really *was* Rampa. He wrote 12 more books as Rampa, before he and Rampa's new body died in 1981. (I wonder whatever happened to Cyril's soul when Rampa's spirit moved in?)

References

Kelly, Lynne, *The Skeptic's Guide to the Paranormal*, Sydney: Allen & Unwin, 2004, pp 83–101.

Welkos, Robert W., 'Giving more life to Marilyn', *Los Angeles Times*, 2 August 2006.

Women Speak More Than Men

(It's All Talk)

A statement said often enough can take on a life of its own. How about this one? 'Women speak 20 000 words each day, but men speak only 7000.' It fits comfortably with all the clichés that television sitcoms and popular pseudo-psychology books spread so effortlessly. You hear it all the time. And it has zero truth.

History of the Myth
– Part 1

Mark Liberman, Professor of Phonetics at the University of Pennsylvania, made it his personal duty to check this oft quoted 'fact' for accuracy. He has worked back through time, trying to nail who said what and when.

The first quoting of this factoid seems to date back to 1993 in James Dobson's book *Love for a Lifetime*. This book, full of Bible-based insights, claims (with absolutely no evidence) that women speak 50 000 words per day, while men can manage only 25 000. In the same year, a marriage counsellor, Gary Smalley, wrote a pamphlet entitled *Connecting with Your Husband*, in which he stated that the average male uses about 12 000 words per day while a woman averages 25 000 words per day.

One factor that immediately made me suspicious was that while the claim was the same — women use more words than men — the number of words attributed to each gender was different. Disparities like this make me wonder where they get their data from.

Allan and Barbara Pease, authors of the 2004 book *Why Men Don't Have a Clue and Women Always Need More Shoes*, said in a CNN interview: '… women can speak 20 000–24 000 words a day versus a man's top end of 7000–10 000.'

Again, women speak more but the numbers are different.

History of the Myth – Part 2

And this was the case with all the references that Professor Liberman tracked down. Women supposedly spoke more, but the numbers of words quoted as being spoken were always different.

Hara Estroff Marano, writing in *Psychology Today*, claimed that '… the average woman uses 7000 words a day and five tones of speech' while 'the average man uses 2000 words and three tones …'.

Kate White, Editor in Chief of *Cosmopolitan*, was quoted as saying that 'the average guy speaks 2000–4000 words a day and the average woman 6000–8000'.

Ladies Golf Journey carried an article by Debbie Waitkus, in which she maintained that 'women speak an average of 30 000 words a day … an average of 12 000 a day for men'.

And Louann Brizendine, in her 2006 book *The Female Brain*, produced yet another set of figures — 'a woman uses about 20 000 words per day while a man uses about 7000'. This book was roundly criticised in the science journal *Nature*. The reviewers of her book wrote: '*The Female Brain* disappointingly fails to meet even the most basic standards of scientific accuracy and balance. The book is riddled with scientific errors …' and 'The text is rife with "facts" that do not exist in the supporting references'.

To Louann Brizendine's credit, when this criticism of her word-count numbers was brought to her attention, she admitted that she had relied on the advice of others and promised to remove these numbers from future editions of her book.

Jaw Problems?

In 2001, Allan and Barbara Pease wrote their pop pseudo-psychology gender-difference book *Why Men Don't Listen and Women Can't Read Maps*. Having compiled a list of different ways that women can communicate the authors finish off with a fascinating observation:

'A woman can effortlessly speak an average of 6000–8000 words a day' with 'an additional 2000–3000 vocal sounds, as well as 8000–10 000 facial expressions, head movements and other body language signals ... a daily average of more than 20 000 communications'. From this they conclude that 'women are four times more likely to suffer from jaw problems'.

Why Women Speak More?

Various 'theories' have been put forward to explain this purported difference between the speaking output of women compared to men.

One particularly ludicrous explanation is that men, as hunters, had to learn to stand still silently for hours while tracking their next meal, while women, as gatherers, would talk incessantly with their fellow female gatherers as they picked berries. But, from what we see of hunter-gatherers in the past and present, hunting took up very little time and both genders gathered fruits and vegetables.

The Actual Research

However, the research that has been done tells us that there is no gender difference in the number of words spoken.

This was known back in 1993. Deborah James and Janice Drakich wrote in their article 'Understanding gender differences in amount of talk: a critical review of research': 'It is shown that the

You say 'I talk to much' like it's a bad thing

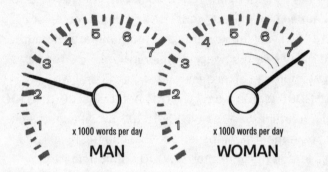

x 1000 words per day
MAN

x 1000 words per day
WOMAN

widely held belief that women talk more than men is unsupported in the literature.'

In 2006, Janet Hyde, a psychologist from the University of Wisconsin, wrote a paper called 'The gender similarities hypothesis'. This was in direct response to the many books on pop pseudo-psychology that claim there are major psychological differences between the genders. These books assert that there are so many differences between how men and women speak, that they may as well be different species. However, as a psychologist (not an author, prepared to write *anything* to get a wide reading audience), Dr Hyde simply had not seen these major gender differences. She analysed 47 peer-reviewed psychological surveys. Overall, women seemed to be 0.11% more talkative than men — a number that is not even statistically significant.

Further research in 2007 supports this view. Over six years, Matthias R. Mehl and his team wired up six separate groups of 17- to 30-year-old university students — five in Texas and one in Mexico. There were 396 students in all — 210 females and 186 males. The tape recorders monitored them for periods of 4–10 days. The results were so close as to be statistically identical — women spoke 16 215 words against the men's average of 15 669. This 500-word difference was tiny compared to the difference between the most talkative and least talkative males (45 000 words).

And yes, the most talkative person was a male, with 47 000 words.

Sexual Word Budget

One study from 1997 looked at the different words that the genders used. The researchers gave Walkman sound recorders to 75 females and 73 males and 'recorded linguistic transactions in which they were engaged during a period of two days'. By including the people to whom their volunteers spoke, they increased the numbers of speakers to 1097.

They found that the words 'most characteristic of male speech' in the larger group that they now surveyed were 'f**king, er, the, yeah, aye, right, hundred, f**k', while the women's most popular eight words were, 'was she, her, said, n't, I, and, to, cos'.

Are They Telling Little Porkies?

The research that has been done clearly shows that with regard to the number of words spoken by men and women, there's no difference — or if there is a difference, it's microscopic. But the popular pseudo-psychology 'gender-difference and relationship self-help' genre of books claim that women speak at least twice as many words as men.

So, just where did the authors of these books get their information from?

References

Brizendine, Louann, *The Female Brain*, Broadway (an Imprint of Random House), 2006.

Dobson, James, *Love for a Lifetime: Building a Marriage That Will Go the Distance*, Multnomah Publishers, 1993.

Hyde, Janet, 'The gender similarities hypothesis', *American Psychologist*, September 2005, pp 581–592.

James, Deborah and Drakich, Janice, 'Understanding gender differences in amount of talk: a critical review of research', in Deborah Tannen (Ed.), *Gender and Conversational Interaction*, Oxford University Press, 1993, pp 281–312.

Liberman, Mark, 'Language log: neuroscience in the service of sexual stereotypes, sex-linked lexical budgets', http://itre.cis.upenn.edu/~myl/languagelog/archives/003420.html.

Mehl, M.R., 'Are women really more talkative than men?' *Science*, 6 July 2007, p 82.

Pease, Allan and Pease, Barbara, *Why Men Don't Have a Clue and Women Always Need More Shoes: The Ultimate Guide to the Opposite Sex*, Broadway (an Imprint of Random House), 2004.

Pease, Allan and Pease, Barbara, *Why Men Don't Listen and Women Can't Read Maps: How We're Different and What to Do About It*, Broadway (an Imprint of Random House), 2001.

Rayson, P., Leech, G. and Hodges, M., 'Social differentiation in the use of English vocabulary: some analyses of the conversational component of the British National Corpus', *International Journal of Corpus Linguistics*, 1997, 2(1), pp 133–152.

Young, Rebecca M. and Balaban, Evan, 'Psychoneuroindoctrinology', *Nature*, 12 October 2006, p 634.

Would You Like Dressing on That Wound?

We all have blind spots in our view of the world. I used to be a doctor at the Children's Hospital in Sydney. But when my kids injured themselves leaving a simple skin wound, I would just clean it and leave it open to the air to 'breathe', thinking that this was the right thing to do. But I was very wrong — wounds do need a dressing.

The word 'wound' comes from the Latin word *vulnus*, which also gives us the word 'vulnerability'. The word 'dressing' comes from the French word *drecier*, meaning 'to arrange', or 'to set right or straight', or 'to put in proper order'.

Part of the difficulty of understanding 'wound healing' is that skin is not made of 'simple' tissues such as fat, fascia (connective tissue) and epithelium (thin layers of surface cells). No, our skin is our personal spacesuit that constitutes 15% of our body weight. It is a very complex organ incorporating many different structures that come from many different embryological origins.

Natural Repair of a Wound

Immediately after a cut in the skin, a blood clot and debris fill the gap. Over the first few days, there is local inflammation and migration of surface (epithelial) cells across the narrow gap. On

Let it breathe ... let it breathe. Actually, cover it up!

Shown below is a simplified example of how healing happens in a clean wound with good contact between the two sides of the wound (the sort of thing that would happen in a surgical cut). Wound healing is more complicated where the cut is not sharp and where there is a gap between surfaces.

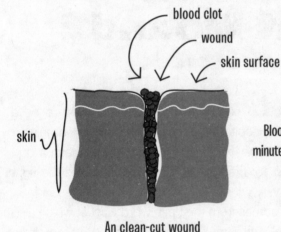

blood clot
wound
skin surface
skin

Blood fills the gap and within minutes turns into a blood clot.

An clean-cut wound

2-3 days

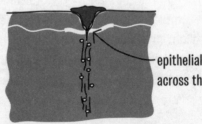

epithelial cells migrate across the wound

The Macrophages eat and remove clot.
Fibroblasts move in and start joining the edges of the wound.

Epithelium

If you are interested in medical stories, you'll keep hearing the word 'epithelium'. The word comes from two Greek roots – *epi* meaning 'above' and *thele* meaning 'teat'. Epithelium is the thin layer of tissue that lines the outside of the body. Human beings have the topological shape of a doughnut, i.e. there is a hole running right through the body from the mouth to the anus, with epithelial tissue lining the entire gut. It also lines the lungs and the urinary and reproductive tracts.

Epithelial tissue is made from cells. In the skin epithelium, the bottom layer of cells looks like tiny columns, the middle layer like tiny cubes and the top layer like flat cobblestones. The top layer of cells contains keratin, a protein that makes the skin tough and relatively impermeable.

Healing a wound has to include healing the epithelium.

about the third day, the blood vessels invade as tiny 'capillary sprouts', moving into the wound at up to 2 mm per day. In about 3–5 days, immune system cells (macrophages) start removing the clot and a protective scab starts to form — thanks to invading fibroblast cells. After about 10–14 days, the scab becomes loose. The epithelial surface skin cells have joined up, the fibrous tissue holding the wound together, although it is still weak. It takes months, perhaps years, to get back to near full strength and a minimal scar.

The scab acts as a semipermeable membrane. It provides a moist environment that encourages the epithelial surface skin cells to migrate across the gap and join up.

Despite the formation of a scab, you still get a better result if you cover the wound. Sometimes, a protective scab doesn't form, perhaps through infection, or the gap being too wide, or the scab is

Let the healing begin ...

10-14 days

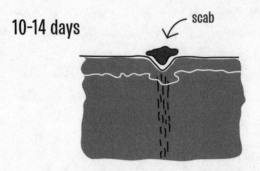

scab

Epithelium fully covers the old wound site.
The scab loosens and falls off.
The edges of the wound are joined with
fibrous tissue, but the join is still quite weak.

Weeks to Months

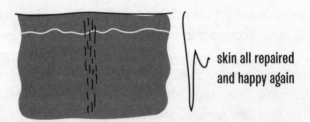

skin all repaired
and happy again

Epithelium is once again almost identical to original.
The scar is still obvious, but will merge into adjacent tissue.
The wound is much stronger, but will take years to get full strength.

accidentally knocked off. Scab formation also depends on the type of wound — whether it's clean or contaminated, and whether it's tidy (e.g. a clean cut from a Stanley knife) or untidy (e.g. a hammer wound).

It is very difficult for us to try to mimic the magic that the body does naturally with the 'scab'. And people have been trying for at least 7000 years to work out how best to treat wounds. In 1995, there were over 1000 different types of wound dressings available, with annual worldwide sales of more than US$350 million.

Scarless Healing

A scar forms after a wound has healed. Although effective, a scar is a poor substitute compared to the original. For example, it's not as strong as skin and, if damaged, doesn't heal as well.

Every wound leaves a scar unless the wound occurs in a foetus in the second or early in the third trimester. We still do not understand why this is so, but we have some ideas.

For example, the foetus has many different growth factors, some of which seem to help repair wounds. One factor is the amniotic fluid, which is loaded with many chemicals. Another is the fact that the foetus is markedly low in oxygen. The blood profile of the foetus (inside the uterus) is very different from that of newborn babies who breathe air through the lungs.

This is why researchers around the world are trying to discover why a wound to a foetus does not leave a scar.

Pare, Poison and Bezoar

In the Middle Ages, the 'bezoar' was thought to be able to neutralise all poisons – it was the Universal Antidote. The name 'bezoar' comes from the Persian word *padzhar*, meaning 'protection from poison'.

Physically, a bezoar was just a lump or 'stone' found in the gut of animals, usually ruminants (animals that have several stomachs), but sometimes in other animals such as human beings. I've seen my cat cough up a furry 'concretion' – yep, a genuine bezoar.

In the Middle Ages, courts were full of intrigue. The nobility would have a small polished bezoar mounted on a ring, so that they could discretely dip it into a drink to supposedly neutralise any poisons that it might contain.

Ambroise Pare tested the effectiveness of the bezoar. A cook at his court stole the silver cutlery. As a 'plea bargain', the cook agreed to take poison and be treated with a bezoar. Of course, the bezoar didn't work and he died in great pain several days later.

On the other hand, the sample size of this experiment was very small – one.

In the TV hospital drama 'Gray's Anatomy', a patient made his own bezoar. In the episode called 'Begin the Begin', one patient was an author who had written a novel. On rereading his novel, he decided that because it was so awful he would literally eat it so that he could 'put it behind him'. It concreted in his gut, forming into a bezoar, which would not pass further and gave him a nasty bellyache.

History of Wound Care – Part 1

The Smith Egyptian papyrus of 1615 BC describes the potions that the ancient Egyptians used to heal wounds. One popular technique involved a combination of animal grease (which doesn't stick — technically called a 'non-adherent'), honey (an osmotic agent and antibacterial chemical) and lint (to absorb secretions — technically called an 'exudate absorber'). By the way, in some wounds, a mixture of one-third honey and two-thirds animal fat can drop the bacterial count of *Staphylococcus* and *E. coli* infections from 100 000 down to 100 in only 24 hours.

A Sumerian clay tablet dating back to 2100 BC gives the three traditional 'healing gestures' — washing the wound (with beer and hot water); making plasters (a mixture of ointments, herbs and oils); and bandaging the wound.

In 1800 BC, Homer's *The Odyssey* tells us how 'with bandage firm, Ulysses' knee they bound'. But around 400 BC, Hippocrates did not like to cover wounds. He used a tin pipe to drain pus out of a wound.

And in 280 BC, an unknown Greek barber invented the first known syringe to suck pus out of wounds.

In the Bible, *Luke* 10: 34 tells the story of the Good Samaritan: 'He went to him and bandaged his wounds, pouring on oil and wine.' Modern science has discovered that the antiseptic chemicals in wine include the pigments malvoside and oenoside. And yes, these are more plentiful in red wine.

The early Greek physician Galen (129–200 AD) discussed various ways of caring for different wounds. He learnt much about this topic during his three or four years as a doctor in a gladiator school. He called wounds 'windows into the body'.

History of Wound Care – Part 2

In the 1500s, Ambroise Pare (1510–1590) — the official royal surgeon to several French kings including Henri II, François II, Charles IX and Henri III — tried to set up a rational system of wound care.

In the 1800s, Dr Joseph Lister (1827–1912) was the first to use antiseptics in surgery. In 1867, he invented the first antiseptic dressing by soaking lint and gauze in carbolic acid. Robert Wood Johnson (a founder of Johnson & Johnson) heard Lister lecture on wound care. In 1891, Johnson & Johnson became the first company to mass-produce sterile, antiseptic wound dressings.

Between 1948 and 1951 various doctors explored the 'moist chamber effect' by covering wounds with various materials including adhesive tapes, permeable and non-permeable nylon films and transparent plastic films. In 1958, Dr Odland discovered that wounds covered by blisters healed better and more quickly when the blister was intact rather than broken. It began to look as though moist wounds healed more quickly and more effectively.

In 1962, the British doctor George Winter, now famous as the Father of Moist Wound Healing, finally proved that wounds that were kept moist healed better than those that were exposed to the air. In technical terms, a moist environment promotes 'epithelialisation', i.e. the joining up of the surface skin cells.

Perfect Dressing

We now know enough to make a reasonable guess as to the properties of the 'perfect' dressing. It should be able to remove toxic chemicals and secretions while keeping the wound humid. It should allow gases to enter and leave but still provide thermal insulation. It should also protect the wound from infection. Finally, it should be easily removed. There is no single dressing that has all six of these properties.

We have learnt more about wound care in the past two decades than in the previous two millennia. I now believe that wounds should be dressed for healing success. One less blind spot for me …

Humble Pare

Ambroise Pare famously said 'I dressed the wound and God healed it'. Working in a time when medical doctors had far fewer effective remedies available, he also said, 'Cure occasionally, relieve often, console always'.

References

Al Ladin, Daniel, 'Understanding dressing', *Clinics in Plastic Surgery*, July 1998, pp 433–441.

Broughton, George, et al., 'A brief history of wound care', *Plastic and Reconstructive Surgery*, June Supplement 2006, pp 6S–11S.

Broughton, George, et al., 'The basic science of wound healing', *Plastic and Reconstructive Surgery*, June Supplement 2006, pp 12S–34S.

Eaglstein, William H., 'Moist wound healing with occlusive dressings: a clinical focus', *Dermatologic Surgery*, February 2001, pp 175–181.

'Healing (Repair) and Hypertrophy', in J.R. Anderson (Ed), *Muir's Textbook of Pathology*, London: Edward Arnold Publishers, 1980, pp 77–101.

Mackool, Richard J., 'Scarless healing: the fetal wound', *Clinics in Plastic Surgery*, July 1998, pp 357–368.

O'Connor, Anahad, 'The claim: wounds heal better when exposed to air', *The New York Times*, 1 August 2006.

Hashish Assassin

Way back in the last century, in the Era of Free Love and Lava Lamps, in the Swinging Sixties and Seventies, marijuana use increased enormously in Western society. One variety of marijuana, hashish, came with its own romantic legend. The exotic gossip spread around at the time suggested that the word 'assassin' came from the Arabic word *haschishin* (meaning 'a user of hashish') and was derived from the word *hashish* (as you might expect).

But no, it's not all true.

The Legend

Westerners in the 1960s and 1970s imagined the Middle East to be a mystical place. And the legend of the origin of the word 'assassin' was true to this idea.

The story goes that an ancient Arabian ruler tricked courageous young men into believing that if they worked for him but died in his service, they would go straight to Paradise. He 'proved' this with a clever ruse, by showing them that he had access to Paradise.

He would, so the legend goes, drug them with hashish, which put them to sleep for three days. After this long and supposedly refreshing sleep, they would wake up in a walled garden surrounded by young women and men of incredible beauty, who would fulfill their every desire. After a few days of total bliss, they would be again drugged with hashish. But when they woke up and returned to their dull lives, they had a very favourable memory of Paradise burned into their psyche. They would then do anything for the ruler and happily undergo training in the martial arts in order to be used as suicide assassins.

This story has some seeds of truth — the best stories always do. The background is complicated, because it involves religion and power. How familiar …

The Religious Reality

The assassin story involves three separate unfriendly religious groups and the leader of one of them, Hassan ibn-al-Sabbah, was also known as 'The Old Man of the Mountains'.

Hassan ibn-al-Sabbah was a leader of the Ismaili religious group — a very small subgroup of Muslims. In the 8th century, the Ismailis had split from the Shiites who were themselves a minority within the Islamic faith.

In 1090, Hassan ibn-al-Sabbah, the charismatic Ismaili leader, seized control of the mighty hill fortress of Alamut in northern Persia with the help of converted Ismaili soldiers inside the castle. In 1094, the Ismaili sect split into the Mustalians (based in Cairo) and the Nizalis (based in Persia), making the Nizalis an even smaller minority in the Islam faith.

To complicate the politics, the Christian Crusaders set off from Europe in 1096 in successive waves to reclaim Jerusalem. They first made contact with the Nizalis in the early 1100s.

So the Christians, the Nizalis and the rest of Islam were locked into a triangle of hostility.

The Assassins

Hassan ibn-al-Sabbah, the Nizali leader, did indeed train young men as *fidais* or assassins.

The assassins' job was to secretly infiltrate noble families and armies as courtiers and soldiers or to travel the countryside as itinerant peddlers and tailors. (In the modern spy trade, they are known as 'sleepers'.) Then, after days, months or years of waiting, they would be 'awoken'. They would kill their target with surgical precision and ruthless efficiency. And if their escape was foiled they would commit suicide in a spectacular act of self-sacrifice. The vast majority of the victims were fellow Muslims, with fewer than eight Europeans being assassinated.

Bernard Lewis, the author of *The Assassins — a Radical Sect in Islam*, describes them as the 'first terrorists'. The Nizalis were intellectuals, excelling in the fields of theology, philosophy, architecture and science. Their warfare was carried out with pinpoint precision aimed at key opponents, not random slaughter of all and sundry. However, their opponents demonised them as bloodthirsty, extremist religious murderers, who ate pork and consumed wine and hashish.

Hashish + assassin = a romantic legend

The Arabic word 'haschishin', meaning 'a user of hashish', came from the word 'hashish' (as you might expect), but has nothing to do with the word 'assassin'.

What a hashish assassin might look like (but then again ... might not look like at all).

The returning Crusaders brought back the legend of the pothead assassins. They loved to believe imaginative, romantic tales of the East and were also impressed by the spectacular self-sacrifice of the Nizalis. However, in 1256, Genghis Khan's grandson, Hulagu, destroyed the Nizali mountain castles, one at a time. He didn't want to be assassinated himself. The political and military power of the Nizalis was permanently broken. However, even today, several million Nizalis still survive in about 25 countries across Asia, Africa, Europe and North America.

The Legend Lives On

In 1273, Marco Polo visited Alamut and brought back the story of how hashish was used to seduce potential killers. Shortly afterwards, Dante (1265–1321) was the first to use the word 'assassin' in canto XXIX of *Inferno* (Hell) in his epic poem, *The Divine Comedy*.

By the mid-1300s, the word had entered various European languages. Today, the word 'assassin' means 'murderer'.

In 1818, the orientalist Silvestre de Sacy published *Dynasty of the Assassins*. He recognised that the assassinations and suicides arose from the Nizalis' complex religious beliefs and practices — but in the context of their belief that they were a righteous minority fighting powerful tyrants. In other words, their assassinations allowed a small group to even the odds against a much larger foe. Even Dan Brown, author of *The Da Vinci Code*, mentions the assassins in his book *Angels and Demons*.

It is an evocative story.

Wrong Again

But there are a few reasons why the link between the words 'hashish' and 'assassin' is almost certainly wrong.

First, if hashish is given in a dose large enough to cause unconsciousness, it will first cause nausea and hallucinations. These are usually very scary and unpleasant side effects for the naïve user, and would not be likely to engender a peaceful sleep, let alone loyalty.

Second, Hassan ibn-al-Sabbah was a deeply puritanical and ascetic person who strictly followed the Koran. Indeed, when two of his own sons were separately accused of drunkenness and murder, he personally had them executed. And he was very much against the use of hashish. He believed that the Koran's ban on alcohol was really a ban on all intoxicating substances.

So, did the word 'assassin' come from the Arabic word *haschishin*, meaning 'a user of hashish'? Almost certainly not. A far more reasonable explanation is that the word 'assassin' came from the leader's name, Hassan. The word *aschishin* means 'follower of Hassan'.

Perhaps it was the other drug-addled, bumbling assassins, whose plans were foiled, that were remembered as having made a hash of it …

Marijuana

Marijuana has been used, both medically and non-medically since at least 2700 BC.

Traditionally, there are three grades. The lowest grade, bhang, is the leaves and stems of the plant, *Cannabis sativa*. The second grade, ganga, comes from the tops of cultivated plants. The top grade, hashish, comes in the form of hard blocks. It is made by scraping resin from the leaves of cultivated plants.

References

Daftary, Farhad, *The Assassin Legends: Myths of the Isma'ilis*, London: I.B. Tauris & Co. Ltd, 1995, pp 1–8.

Lewis, Bernard, The Assassins; *A Radical Sect in Islam*, London: Weidenfeld & Nicholson, 2003.

'The Hashashin', *Fortean Times*, August 2006, p 50.

'The straight dope: does "assassin" derive from "hashish"', http://www.straightdope.com/mailbag/massassin.html, 30 July 2002.

The Lion's the Mane Man

The first thing you notice about a lion is his mane. Most people think that the mane is a simple and obvious signal to the female lion or lioness. They believe that the mane advertises that the possessor of the mane will give the lioness excellent lion cubs, with the added bonus of being a good provider and protector.

However, according to zoologists, the purpose of the mane is still a big mystery.

Lion 101

The lion is a large, well-muscled cat that can grow to 2 m in length, with a shoulder height of 1.2 m, and weigh nearly one-quarter of a tonne. The female is smaller. Lions reach their reproductive prime at around age 5–7 years of age. Wild lions breed no more than once every two years. As the female is only 'available' for mating for 3–4 days, the male will mate with her up to 50 times per day during this very brief window of opportunity. Wild lions are usually dead by the age of 10 but can live up to 25 years in captivity.

From about two million to 10 000 years ago, lions ranged freely across North America, Africa, the Balkans, the Middle East and India. They vanished from North America 10 000 years ago, from the Balkans 2000 years ago and from the Middle East around

800 years ago. By 1980 there were approximately 100 000 lions in the wild, but today there are only about 25 000. Human beings, quite appallingly, have slaughtered about three-quarters of the lion population in just 25 years. About 200 lions remain in India, with the rest in Africa, south of the Sahara.

The other unique feature of the lion is its sociability. They are the only members of the cat family that live together in groups. We still don't know why this is so, but there are lots of theories.

A group of lions, called a pride, averages about 15 members but can range from four to as many as 37. The pride consists of several generations of lionesses (usually related), cubs and a few males. When new males successfully take over a pride by fighting and forcing out the existing males, they will usually kill the cubs, forcing the females to mate again.

The pride's territory varies between 20 and 400 km^2, depending on the density of the available prey. The territory usually has a well-defended core, surrounded by a fringe area where neighbouring prides and other predators are tolerated.

Big Eater

Many of the lion's hunts fail. Lions often allow the wind to carry their scent onto their prey. And they cannot run long distances after their prey, because they get tired – they are better sprinters than marathon runners.

However, they have a huge appetite. A hungry adult male can eat 34 kg of flesh in a single meal – and then not eat for a week.

If there is a lot of food around, they will hunt for only 2-3 hours per day, spending the remaining hours resting, sitting and sleeping.

So Why the Mane?

This question was first posed by Charles Darwin. He believed that it might be a sign of sexual vigour or perhaps a defence against other lions. A large bushy mane might make it difficult to see and attack the neck.

The question is hard to answer because lions are notoriously hard to study. First, they are nocturnal animals. Second, they inhabit the same range as other aggressive animals (such as elephants) that also do not tolerate human beings.

Robust lion research probably began in the 1960s with George Schaller. He studied the lions of the Serengeti National Park, which covers about 15000 km^2 of Tanzania in Africa. Because of its altitude the park has a relatively cool climate for an area so close to the Equator.

The mane usually develops at puberty. For Serengeti lions, it begins to develop at about one year of age and continues to grow, getting darker as they age. However, the size of the mane is highly variable, both from one population of lions to another and within any given population. In size, a mane can vary from a tiny 'mohawk' on the top of the head with very scant side whiskers right up to a very thick coat covering the chest and shoulders.

The mane, impressive as it is, has a few disadvantages for the lion. It takes energy to grow and maintain a mane, which can then hide and shelter parasites. A sizable mane will also reveal the lion's location to its prey and decrease his ability to manoeuvre through thick bramble and undergrowth.

Acting like a woollen scarf, the mane retains heat. It is difficult for lions to lose heat because they cannot sweat and do not have a cool, wet nose. They can only lose heat from their skin or by panting. In fact, on hot days they have been observed scraping away the warm surface dirt, so that they can lie down in the cooler dirt underneath.

Cool Old Cats Lead Mane Event

One study looked at 19 lions in 17 zoos in various climates across the USA. It found that the local temperature was responsible for up

The main thing about the Lion's mane is ...

The lion's mane

The question of why the lion has a mane was first asked by Charles Darwin.
He thought that it might be a sign of sexual vigour,
or perhaps a defence against other lions.
The question is hard to answer because lions are notoriously hard to study.
First, they are nocturnal animals. Second, they live in the same territory as
other aggressive animals ... and finally ... they don't really like humans.

A (simplified) human.
Lions don't generally
tolerate humans.

to half of the length of a lion's mane. If the local temperature was high, the mane was small and vice versa. The remaining causes of mane size included nutrition, genes and social factors.

However, there is a problem with this study. Most captive lions have unknown pedigrees, so it is impossible to estimate their 'natural' mane length (i.e. as it relates to the mane length of their fathers and of lions in the area where they were born). In addition, captive lions can be subject to external stresses, including chronic inactivity and climate-controlled environments.

All in all, captive lions are not the best animals to study.

A second study, carried out over nine years, looked at two separate populations of lions in the wild — with similar results. Of the lions studied, 78 inhabited the Serengeti National Park at an altitude greater than 1400 m) and 55 inhabited the Tsavo National Park in Kenya at an altitude less than 800 m). The parks were only a few hundred kilometres apart. However, because they were at different altitudes, they experienced different temperatures, with Tsavo hotter and more humid.

The study found that the size and darkness of the mane was mostly related to the age of the lion and the local temperature and humidity. The mane was large if the temperature and humidity were low, and the lion was old. At their sexual peak, lions in the warmer Tsavo National Park had quite small manes that continued to grow with age. The study also found that there was no link between a large impressive mane and sexual maturity. In fact the author said, 'usually lions are well past their breeding prime when they carry the most extensive and often darkest manes of their lives'.

So if you're a male lion with a large mane, thank the cold and being old. And if you really want to impress a lioness, grow up in a cold climate and then move, with your large mane, to a warmer climate where all the local males have small manes.

And if you're a lion with a small mane, you might like to seek solace with the man who has small feet — there's no correlation in either case with sexual vigour.

Ghost and Darkness

Tsavo National Park in Kenya is huge, covering approximately 20 720 km². It ranges from the arid regions in the east to the green hills in the west.

In the late 1890s, two lions (8-10 years of age) terrorised the workers laying tracks for a new railway line. They killed about 135 Indian and African railway workers, completely halting the progress of what is now the Kenyan Railway Line. Lt Col J.H. Patterson killed them in 1898. Their bodies were stuffed and are now housed in the Field Museum in Chicago.

In 1996, Michael Douglas and Val Kilmer starred in a movie about this event called *The Ghost and the Darkness*.

References

Caputo, Philip, 'Maneless lions', *National Geographic* (Online Extra), http://magma.nationalgeographic.com/ngm/0204/feature2/fulltext.html, April 2002.

Encyclopaedia Britannica, Ultimate Reference Suite DVD, 2006 — 'lion'.

Gnoske, T.P., et al., 'Dissociation between mane development and sexual maturity in lions *(Panthera Leo)*: Solution to the Tsavo riddle', *Journal of Zoology*, December 2006, pp 551–560.

Hill, Geoffrey E., et al., 'Melanin, nutrition and the lion's mane', *Science*, 31 January 2003, p 660.

Kays, Roland W., et al., 'Mane variation in African lions and its social correlates', *Canadian Journal of Zoology*, Vol 80, 2002, pp 471–478.

Patterson, B.D., et al., 'Developmental effects of climate of the lion's mane *(Panthera Leo)*', *Journal of Mammalogy*, Vol 87(2), 2006, pp 193–200.

West, Peyton M., 'The lion's mane', *American Scientist*, May/June 2005, pp 226–236.

West, Peyton M., et al., 'Sexual selection, temperature and the lion's mane', *Science*, 23 August 2003, pp 1339–1343.

Withgott, Jay, 'Cool cats lose out in mane event', *Science*, 23 August 2003, pp 1256, 1257.

Mobile Phone Cooks Egg
(Poaching the Truth)

Recently my email inbox began filling up with the latest 'terri-frying' story about the dangers of mobile phones. The attachment 'Microwavinganegg' claimed that the radiation from a mobile phone placed near an egg would actually cook the proteins in the egg. The email then put forward the chilling proposition, 'Imagine what it can do to the proteins in our brains...'

The Instructions

The email came with quite detailed instructions. All you have to do is place an egg in an egg cup, sandwich it tightly between two mobile phones facing each other and then get one phone to call the other phone for 65 minutes. It even provided a timetable. It says that nothing much happens for the first 15 minutes. The egg then begins to get warm in about 25 minutes, is hot to the touch in 45 minutes and is actually fully cooked and ready to eat in 65 minutes.

The email attachment did not identify the author but it was definitely not linked to any university or school. It was full of bad grammar, misplaced apostrophes and atrocious punctuation (all of them capital sins in my book).

But what if it was possible? After all, regular infrared radiation can toast a slice of bread.

The Experiment

The experiment was ridiculously easy to set up. My eight-year-old daughter, Lola, and I began by drawing up a chart on a sheet of paper. Then we placed one phone on either side of the egg. We recorded the data, detailing the timing of our phone calls. We made the first of seven 10-minute calls — on our mobile phone plan, the first 10 minutes of each call are free — and let the experiment run.

After 40 minutes my daughter was soooooo bored by the complete lack of any heating action of the egg, that she gave up colouring in our results sheet. After 70 minutes the egg was still at room temperature — and the insides were still runny.

I did the whole experiment again the next day with an accurate thermometer. After 70 minutes sandwiched tightly between two operating mobile phones, the temperature of the egg had not budged by a single degree — and the egg was definitely not cooked.

It was not surprising if you look at the physics.

Lola's chart

Eggs, keypad side up!

According to urban myth, you can cook an egg placed between two
mobile phones that have 'called' each other.

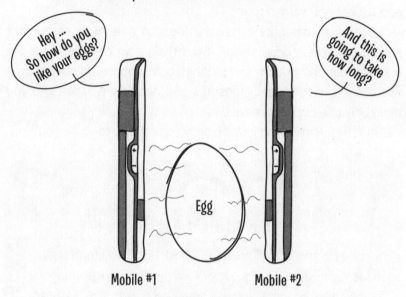

Hey ... So how do you like your eggs?

And this is going to take how long?

Mobile #1   Mobile #2

Egg

The Physics

First, consider the power needed to cook an egg. Mobile phones
transmit a maximum of 2 watts of peak power. But this maximum
power is transmitted in short pulses for only one-eighth of the time.
When you average the power transmitted by the phone over a
second or more, you get an average power of 0.25 watts, or one-
quarter of a watt.

However, this peak power level is used only when the phone is
at its maximum distance from the base station. When it's close to a
base station, the peak power that the phone transmits can drop to
as little as 2 milliwatts (two-thousandths of a watt). That's a
reduction of 1000 times. This is a lot less than the 600–1000 watts
of a typical microwave oven.

If the two mobile phones were transmitting at their maximum
power and all of their power went only into the egg, it would take

about 40 hours to cook it — provided that the egg did not lose any heat into the local environment. But the egg would lose these tiny amounts of heat as fast as it gained them, which explains why my eggs did not get warm. In fact, the British TV show 'Brainiac' did this experiment using 100 phones and got the same result — the egg did not get warm.

Second, mobile phones do not transmit directly to each other but to and from a base station. So it makes no sense to have the egg between the phones. And further, both phones do not transmit at the same time — one transmits and one listens. Therefore you need only one phone, not two.

So is this whole thing just a psychological ploy to scare you?

Divide by Eight

Mobile phones do not transmit continuously. Instead they transmit in short bursts, separated by much longer periods of no transmission. If you use your phone near a hi-fi system, you can hear these bursts coming through the speakers.

Each second is broken up into 217 'frames'. Each of these 217 frames is about 4.6 ms long (a millisecond is one-thousandth of a second). Each frame is then further broken down into eight 'slots', each about 0.58 ms long. The mobile phone transmits information only in the first of these eight slots. No information is sent in the remaining seven slots. This is why the average power is one-eighth of the peak power.

Because mobile phones that transmit at 900 MHz have a peak output power of 2 watts, their average power is one-eighth of this – 0.25 watts (or 250 milliwatts or thousandths of a watt). The corresponding powers for mobile phones running at 1800 MHz are 1 watt and 0.125 watts (or 125 milliwatts).

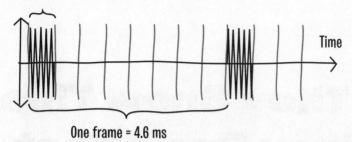

One slot = 0.58 ms

Time

One frame = 4.6 ms

Who Dun It?

Let's look at how it all started.

This myth began with Charlie Ivermee, a 61-year-old archiver at a legal firm in Southhampton in the UK. He has a web page, the *Wymsey Village Web*, that recounts the goings-on in the fictional village of Wymsey. He writes under the name of Suzzanna Decantworthy.

Charles, who has a background in electronics, fessed up to starting this myth. He found the whole concept of mobile phones frying your brain really silly. So in 2000, he wrote a send-up story about cooking eggs with your mobile. He was more than a little surprised to find people actually believed such a silly story.

To me, the Take Home Message is do the experiment and test it for yourself. Believe it when you see it with your own eyes. If an eight-year-old with one egg, two mobile phones and 60 minutes can run the experiment, so can any doubting Thomas.

References

http://www.gelfmagazine.com/mt/archives/how_to_cook_an_egg_and_create_a_viral_sensation.html

http://www.hpa.org.uk/radiation/understand/information_sheets/mobile_telephony/mobile_phones.htm

http://www.mobileshop.org/safety/microwave.htm

http://www.wymsey.co.uk/wymchron/cooking.htm

They Snow This, They Snow That...

Most of us know very little about the finer details of the Eskimo language. But if we are pushed, many of us can drag out from some deep recess of our memory something about 'snow'. That solitary remnant of knowledge is that the Eskimo language has hundreds of words for 'snow', unlike the lexically deficient English language which has only one. This simple little 'fact' is wrong on so many different levels.

Eskimo People

Eskimos, along with the closely related Aleut, make up most of the native population that lives in the Arctic and subarctic regions of Siberia, Greenland, Canada and Alaska. They call themselves by various names, most of which mean 'the real people' or 'the people' in the local language. The people of Greenland and Canada call themselves the 'Inuit', while those in Alaska prefer the name 'Eskimo'.

They have been living in these far northern regions for at least 3000 years. We know this from excavations on Umnak Island in the Aleutian Islands. By the beginning of the 21st century, there were approximately 117 000 people who called themselves Eskimos — 51 000 in Denmark and Greenland, 43 000 in Alaska, 21 000 in Canada and only about 2000 in Siberia.

Eskimos are quite different from North American Indians. Most Eskimos carry the B blood group, while very few (if any) North American Indians have it. Eskimos also 'look' more Asian and tend to have smaller hands and feet.

The languages that they speak — the Eskimo–Aleut languages — are also very different from those spoken by the American Indians.

Aleutian Islands

The Aleutian Islands lie approximately 1800 km in an arc from western Alaska towards eastern Siberia. The islands on the Siberian side are called the Commander Islands. There are about 14 large islands, 55 smaller islands and a huge number of tiny islands – a total surface area of about 17 666 km². The islands form part of the Ring of Fire, the volcanos that surround the Pacific Ocean.

The Aleutian Islands were discovered in 1741 by a Russian exploration team led by the Dane, Vitus Bering and the Russian, Aleksey Chirikov. They became separated in a storm and when it cleared, Bering 'discovered' some of the western islands, while Chirikov 'discovered' some of the eastern ones. (I say 'discovered', because the Aleutians were already living there.)

Some of the islands were home to many fur-bearing animals, which were killed in vast numbers as part of the very lucrative fur trade. In 1867, Russia sold the Aleutian Islands to the USA as part of a package that included Alaska.

In 1942, the Japanese successfully invaded two of the western islands, interned the inhabitants and held the land for a year. American troops booted them out after 19 days of fighting.

So yes, technically speaking, in World War II, the Japanese did invade and hold a piece of American territory.

Eskimo-Aleut Languages

There is no such thing as a single language spoken by all Eskimos, partly because these peoples inhabit such huge expanses of widely separated land.

The Eskimo languages are known as 'polysynthetic' languages, i.e. words can be any length, with the addition of multiple suffixes. Therefore, there is no limit to the number of words in any of the Eskimo languages and, indeed, to the number of potential words for 'snow'.

There are about 75 000 fluent speakers of the Eskimo languages. There are actually two major languages — Inuit (spoken in Greenland, Arctic Canada and northern Alaska) and Yupik (spoken in Siberia and southwest Alaska). These two languages are mutually unintelligible. Even within the Inuit language there are many dialects, varying greatly from place to place. In fact, only people in neighbouring localities can understand each other.

There are about 2000 Aleut people, of whom about 100–200 speak the language fluently. The Aleut language is spoken mainly in the Aleutian Islands, situated between Siberia and Alaska. There are two major dialects, which are mutually intelligible.

So Eskimo peoples speak many different languages or dialects.

History of the Myth – 1911

Geoffrey K. Pullum wrote the book *The Great Eskimo Vocabulary Hoax and Other Irreverent Essays on the Study of Language*. He discussed the myth of the 'multitude of snow descriptors used by these lexically profligate hyperborean nomads, about whom so little information is repeated so often to so many', and showed how little truth there was in it. He based much of his discussion on the work of Professor Laura Martin.

Pullum says that this myth began in a low-key fashion with Franz Boas in 1911.

In his Introduction to *The Handbook of North American Indians*, Boas discussed the concept of 'derived' and 'independent' terms. For example, derived terms from the word 'water' would include

You say snow,
I say snow ... we all say snow

There is a long standing belief circulating that the Eskimo language has
hudreds of words for 'snow'.
This simple fact is wrong on so many different levels.

rainwater, sea water, fresh water, and brackish water, and
independent terms would include lake, ocean, wave, dew and foam.

Boaz then gave some independent Eskimo terms for the word
'snow'. He provided exactly four examples — *aput* for 'snow on the
ground', *qana* for 'falling snow', *piqsirpoq* for 'drifting snow', and
qimuqsuq for 'snowdrift'.

That was it — *four* examples.

History of the Myth – 1940

Benjamin Lee Whorf then enters the story.

Whorf had a Bachelor of Science degree in Chemical
Engineering from the Massachusetts Institute of Technology (MIT).
He worked for the Hartford Fire Insurance Company in Connecticut
as a fire prevention officer.

But Whorf was also a very talented amateur linguist. He single-handedly deciphered Classical Mayan hieroglyphs, opening up to study an entire language that was previously lost. He also invented very useful (and very specific) linguistic terms, such as 'cryptotype', that are used even today. Whorf is deeply respected for his lasting work.

However, making mistakes is part of being human and Whorf's 1940 article, 'Science and Linguistics', showed that he was very human. He wrote the article not for a peer-reviewed linguistic journal, but for MIT's promotional magazine, *Technology Review*. In it he claimed: 'We have the same word for falling snow, snow on the ground, snow packed hard like ice, slushy ice, wind-driven flying snow' whereas 'an Eskimo … would say that falling snow, slushy snow … are sensuously and operationally different. He uses different words for them and for other kinds of snow.'

This single paragraph has two major flaws.

First, Whorf was wrong about the English language. We do have a number of independent words for 'snow'. We call it 'snow' when it is white and fluffy, 'sleet' for partly melted falling snow, 'slush' for muddy melting snow and so on.

Second, with absolutely no evidence nor any specific knowledge of Eskimo languages, he had increased the number of purported Eskimo words for 'snow' to seven.

History of the Myth – 1958+

In 1958, Roger Brown quoted this 'fact' in his book *Words and Things* — and he too got the number of Eskimo words for 'snow' wrong. Geoffry Pullum writes: 'After works like Brown's have picked up Whorf's second-hand misrecollection of Boas to generate third-hand accounts, we begin to get fourth-hand accounts carelessly based on Brown.'

In fact, after Brown, nobody goes back to Boas — the original source. The misquoters multiply like rabbits. Carol Eastman quotes the 'fact' in her *Aspects of Language and Culture* in 1975. By 1978,

Lanford Wilson's play *The Fifth of July* claims that there are 50 Eskimo words for 'snow'.

And 1984 was a bumper year. First, *The New York Times*, a 'journal of record', misquoted Whorf to bump the total number of Eskimo words for 'snow' to 100. Second, the Cleveland TV station WEWS in a weather segment claimed that Eskimos had 200 words for 'snow'. Third, Cecil Adams in his *The Straight Dope: A Compendium of Human Knowledge* gave a wry explanation. He claimed that the Eskimos 'have a limited environment to talk about, so they have to make up a lot of words to fill up their conversations'.

Finally, in 1986, Professor Laura Martin tried to set the record straight with her serious academic paper in the peer-reviewed journal *American Anthropologist*. She discussed the wide circulation of this myth and provided a timeline of its origins and subsequent spread. She called it 'an object lesson on the hazards of superficial scholarship'.

Despite her work, this myth is still firmly entrenched in our oral tradition.

But Really, How Many Words?

Exactly how many Eskimos words are there for 'snow'.

According to Professor Anthony Woodbury, an Eskimologist at the University of Texas at Austin, this depends on your definition of the words 'Eskimo', and 'snow'.

First, 'Eskimo'. As I have already mentioned, there are many separate peoples scattered over huge expanses of land who have the generic name of 'Eskimo'. And they speak different languages — there is no single Eskimo language.

Second, 'snow'. In English, we can generate lots of words from the basic word 'snow', e.g. 'snow goose', 'snow job', 'snow tyre' and so on. It works the same way in any of the Eskimo languages. But these languages have a major difference. They allow for about 280 different inflections or grammatical endings, as well as a huge number of suffixes. There is no limit to the length of an individual Eskimo word or to the number of words that can exist.

Finally, Professor Woodbury draws our attention to the fact that many of the Eskimo words that supposedly relate to 'snow', actually do not. The word *igluksaq* was claimed to mean 'snow for igloo making'. But the word actually comes from two words, *iglu* meaning 'house' and *ksaq* meaning 'building material'. So *igluksaq* simply means 'materials for building a house', and these could possibly include bricks, nails, insulation, windows and perhaps even the cabling needed to join computers into a small network. It is not a word for a specific type of snow.

In 1987, Professor Woodbury was pushed by his students to provide an exact number of Eskimo words for 'snow'. He came up with a dozen or so. He was prepared to concede perhaps a couple of dozen, if you included natural phenomena (such as blizzards), behaviour and events.

The number of words commonly used by Eskimos for 'snow' is few, but the discussions that they can have about its vicissitudes are long and detailed. However, this is not language but culture — snow and ice are a large chunk of their daily lives.

It's All So Silly

First, why do we believe such a myth?

Well, to Westerners, Eskimos are exotic and easy to recognise as a separate group. Professor Martin writes that '... easy generalisations are routine: they eat only raw meat, they give their wives as gifts to strangers, they rub noses instead of kissing, they send their elderly out on ice floes to die. We are prepared to believe almost anything about such an unfamiliar and peculiar group.'

Second, what's the big surprise if they do have hundreds of words for 'snow'? Car mechanics have hundreds of jargon words to describe various aspects of the internal combustion engine e.g. crank, big end, little end, conrod, points and carbie. In fact, every field, ranging from carpentry to music and medicine, has its own special words.

In this myth, the number of Eskimo words for 'snow' just grew and grew, with remarkably little evidence to support it.

Professor Martin finishes her article with words of warning about '... the careful use of sources, the clear presentation of evidence and, above all, the constant evaluation of our assumptions'.

References

Adams, Cecil, *The Straight Dope: A Compendium of Human Knowledge*, Chicago Review Press, 1984.

Boas, Franz, Introduction to *The Handbook of North American Indians*, Smithsonian Institution Bulletin 40, Part 1, 1911 (reissued by the University of Nebraska Press, 1966).

Encyclopaedia Britannica, Ultimate Reference Suite DVD, 2006 — 'Eskimo–Aleut languages'.

Martin, Laura, 'Eskimo words for snow: a case study in the genesis and decay of an anthropological example,' *American Anthropologist*, New Series, Vol 88, No 2 (June 1986), pp 418–423.

Pullum, Geoffrey K., *The Great Eskimo Vocabulary Hoax, and Other Irreverent Essays on The Study of Language*, University of Chicago Press, 1991, pp 159–171.

'There's snow synonym', Editorial, *The New York Times*, 9 February 1984.

Push My Button

We all spend time waiting for an elevator to arrive or waiting at a pedestrian crossing for the traffic lights to turn green. And there's always somebody standing there trying to speed things along by incessantly jabbing their finger on the button.

These people believe that by pressing the button frequently and vigorously the elevator or 'walk' sign will take less time to appear. They're wrong.

Traffic Light Controllers

Several different types of control circuits manage traffic lights.

Some traffic lights in large cities have no external detectors or press buttons at all because of the constant flow of traffic day and night. These traffic lights simply work on timers.

However, in areas where there is less traffic, e.g. in the suburbs or on country roads, a combination of control boxes, car detectors and input switches is used.

A control box has timers, processing circuitry and relays.

Car detectors can range from rubber hoses full of air to super-sensitive lasers. But the most common detection technique today is the inductive loop — coils of wire embedded in the surface of the road. When a car (which has a lot of metal in it) rolls over these coils of wire, it becomes part of the circuit which then tells the control box that a car has arrived. (By the way, some very small motorbikes do not have enough metal to set off the more insensitive inductive loops. This is why you occasionally see riders jumping off their motorbikes to press the 'walk' button.)

The 'walk' button is the most common input switch. However, most city traffic lights can accept an override command from the Traffic Control Centre, e.g., to give a continuous set of green lights for ambulances, fire engines or VIPs.

Press All You Like

In New York City the traffic lights are being increasingly controlled by computers, the buttons are becoming obsolete. Fewer than 20% feed signals to the control boxes while the remaining 80% are empty props.

In Sydney's Central Business District, the buttons have no effect between 7.00 am and 7.00 pm Monday to Wednesday and 7.00 am to 9.00 pm Thursday to Saturday. However, they do work outside these hours, including all day Sunday. The rationale is that during busy periods the road traffic is relatively constant and the pedestrian crossings are in continuous use.

However, Howard Scruby, executive director of the Pedestrian Council of Australia, said that the pedestrian phases were too short and that handicapped and elderly people found it impossible to cross the road before the red light appeared.

Too Cool

A large group of pedestrians accumulate at a traffic light. The first person to arrive doesn't press the button because they don't realise that it's necessary. And the later arrivals don't press the button because they think that it has already been pressed. What will happen?

If the buttons are wired into the control circuit, the lights will not turn green for the pedestrians. So don't be too cool to press.

'Push it, baby ...
push it real good!'

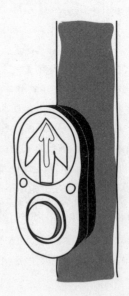

The 'walk' button is the most common input switch. And there is always someone who will try to speed things along by jabbing their finger on the button over and over.

This DOES NOT work!

Timing

You have probably noticed that when you press the Walk button, the lights sometimes change almost immediately, but sometimes there's a delay of a few minutes. The long delays usually occur in rush hour when a timing program kicks in. There is a reason for this. It is better to keep one pedestrian waiting rather than hundreds of cars travelling in or out of town.

However, it is important to note that it takes only one stab of your finger on the button to set the process going. Pressing it once or a hundred times will not make any difference to the control circuit. It's exactly the same with the call button on an elevator — one single press is all that you need. (Traffic authorities in the UK used to have a system that counted these 'button pushes' to regulate the pedestrian waiting time. But it was discontinued.)

In general, people vastly overestimate the amount of time that they spend waiting at traffic lights and elevators. Even if the wait is just a few seconds, it feels like a minute.

Of course, some people claim that the buttons are just there merely to trick the ignorant and taunt the wise, and have no effect at all because they are not even wired into the circuit. They are convinced that the green light or the elevator always follows a timing program and appears when the control program dictates. They reckon that these buttons are just a modern mechanical placebo, designed to placate the hyperactive people in our society.

The Button

During the Cold War, the two most powerful men in the USA and the USSR were the men with 'the finger on the button' – the button that would unleash a nuclear catastrophe on the other side.

Shabbat Elevator = No Buttons

In the book of *Exodus* 31: 12–17, God tells Moses how important the Jewish Sabbath (Shabbat) is: 'Keep the Sabbath as something sacred to you. Anyone doing work shall be cut off spiritually from his people, and therefore, anyone violating it shall be put to death. Do your work during the six week days, but keep Saturday as a Sabbath of sabbaths, holy to God.'

There are 39 activities that are forbidden on the Sabbath. These are broken into the broad categories of baking bread, making garments, making leather and building a house. Number 37 forbids the igniting of a fire, and this is the category that covers elevators. (Elevators run on electricity, and electricity is a kind of fire.)

For this reason Orthodox Jews, i.e. those who adhere strictly to Jewish law, can use only a special Shabbat elevator on the Sabbath. It operates following a preset clock, and is not influenced by the passengers in any way. It stutters its way slowly up and down the building, stopping at each floor for a preset time, and automatically opening and closing the doors without any interventions required by the passengers.

Gambling

Another common human behaviour involving button pressing is playing poker machines. There is a link to rat behaviour experiments here.

A group of psychologists who were studying rats that were pressing levers in the hope of getting food made an interesting finding. When they increased the time of the food delivery, the rats responded by pressing the levers less frequently.

The psychologists believed that this finding might help solve the issue of problem gamblers addicted to poker machines. They suggested increasing the time delay between payouts. However, the gambling industry objected, presumably on the grounds that it would probably work.

References

Brown, Malcolm, 'Little green man has a mind of his own', *The Sydney Morning Herald*, 3 March 2004, p 3.

Dundes, Alan, *The Shabbat Elevator and Other Sabbath Subterfuges: An Unorthodox Essay on Circumventing Custom and Jewish Character*, reviewed by Mikel J. Koven in *Journal of American Folklore*, Washington: Fall 2005, Vol 118, Number 470, pp 501–502.

'Press here to control the universe', *The New York Times*, 1 March 2004.

The Gum and the Restless

You may not realise this, but there's a very clear demarcation line between doctors and dentists. Doctors 'have the run of' or 'own' your entire body — except for the 32 little white things in your mouth, which 'belong' to dentists. Doctors don't know much about teeth. Nor do I — I used to be a medical doctor. So when I went to see my oral surgeon to find out about the pain in my gums, I was very surprised to find that the health of your gums can affect your overall health.

History of Dentistry

The Baltimore College of Dental Surgery — the first 'regular' dental school — opened in 1840 in Maryland in the USA.

But 9000 years ago, there were dentists in Pakistan using drills made of flint. Their knowledge must have been passed from teacher to apprentice, providing some kind of dental teaching.

About 5300 years ago, their counterparts in the Indus Valley knew dentistry, as did their Egyptian counterparts 5000 years ago. The Greeks and the Romans also performed dental surgery — and even made artificial teeth.

Pierre Fauchard, the 17th century French physician, is often called the 'Father of Modern Dentistry'. He maintained that sugars could cause tooth decay and used artificial teeth and fillings for dental caries. He ensured that dentistry was recognised as a real profession, with an unforseen result. The mouth was 'torn away' from the rest of the body, both professionally and philosophically.

Tooth Anatomy 101

The bits of your teeth that you can actually see are the crowns (tops) of the teeth. They are fairly flat. The exteriors of the crowns are covered with the hardest material in the body — enamel. Enamel is dead. It contains no living cells — just like hair. However, immediately underneath the enamel is dentine, a living material similar to bone — but not as hard as enamel.

The bits of your teeth that you can't see are the roots, which reach down below the gumline. Below the gumline, the dentine is covered by cementum, a living material like bone but not as hard as enamel.

In the very centre of the tooth is the pulp, which has various types of cells, a nerve and various blood vessels. The pulp nourishes the dentine, keeping it alive.

And finally, the periodontal ligaments — bundles of fibres — join the roots of the tooth to the bone.

Human beings get two sets of teeth during their lifetime.

The first set (baby, primary or deciduous teeth) start pushing through the gums at around six months of age, are fully 'installed' by two and a half years of age, begin to fall out at 5–6 years of age and have usually all disappeared by 13 years of age. There are usually 20 of them.

The second set usually has 32 so-called permanent teeth. A complicated 'dance' begins as the roots of the primary teeth are dissolved, the primary teeth fall out, and the permanent teeth push towards the surface — but it usually works well.

Teeth are very useful. You can use them for catching and chewing your food and even as a weapon of defence.

The gum and the restless

The bits of your teeth that you can see (the tops) are called the crown.
The outsides of your teeth above the gums are covered with enamel
(the hardest material within your body).

A damn nice smile
with lots of teeth

The anatomy of a tooth

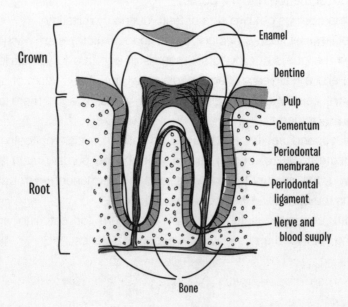

Crown

Root

Enamel

Dentine

Pulp

Cementum

Periodontal
membrane

Periodontal
ligament

Nerve and
blood suuply

Bone

Tooth Anatomy 102

The hardest material in the tooth is enamel. It's 95% inorganic hydroxyapatite crystallites $Ca_{10}(PO_4)_6(OH)_2$. The rest is 4% water and 1% proteins.

Dentine is about 75% hydroxyapatite, 20% organic material (mostly fibrous collagen) and 10% water.

Cementum is about 50% hydroxyapatite and 50% organic collagen matrix and water.

The pulp has many types of cells – mostly fibroblasts, which have the job of forming and maintaining the pulp compartment. There are also immune system cells, as well as vascular, nerve and lymphatic cells.

Tooth Disease

The word 'periodontitis' comes from three separate roots — 'peri' meaning 'around', 'dont' meaning 'tooth' and 'itis' meaning 'inflammation of'. So, in general terms, the word 'periodontitis' means some kind of inflammation of the soft tissues around the teeth.

Periodontitis can start with 'plaque', the sticky film that forms on the teeth. A build-up of plaque in the spaces between the gums and the teeth can lead to gingivitis, a relatively minor inflammation of the gums. About one-third of Americans have persistent gingivitis. At the other end of the spectrum is periodontitis. With this disease, gaps or pockets between the teeth and the gums fill with pus, a mixture of living and dead bacterial cells and your own living and dead immune system cells.

These pockets of pus are sometimes very shallow but can occasionally penetrate deep into the very roots of the tooth. The bacteria can then eat away the bone under the gum, threatening the life of the tooth.

A pocket 1–2 mm deep, with no bleeding, is of little concern. If the pockets are 3–4 mm deep, they may need regular and more thorough dental cleaning. However, if they are 5 mm deep, the tooth is in danger and dental surgery is required. And if the patient is diabetic — or pregnant — more care is needed.

The gums do not usually hurt until some irreversible damage has already been done.

However, periodontitis is not just a local disease — it can affect parts of your body far from the mouth.

Mouth Gateway to Health – Part 1

My oral surgeon previously held a position at a hospital that performed all kinds of transplants — kidney, lung, liver and heart. Each of these transplant operations had its own specific failure rate. But when he treated periodontitis in patients scheduled for heart transplant surgery, the failure rate halved.

The transplant surgeons and my oral surgeon were not really sure why this was so. However, they believed that it was related to these patients having a collection of pus somewhere in the body — in this case, near the roots of the teeth. They guessed that the pus was periodically releasing a shower of bacteria into the bloodstream. This would keep the patient's immune system mopping up the shower of bacteria as best it could.

In transplant patients, the showers of bacteria in the bloodstream could infect the transplanted organ, making it fail. Therefore, removing the pus improved the success rate of these operations. This confirms a saying that I learnt in surgery, 'If there is pus about, let it out' — in other words, remove it.

Mouth Gateway to Health – Part 2

Let's look at how this problem could affect an otherwise healthy population in the long term.

Showers of bacteria from the periodontitis spraying into the bloodstream every few hours over months and years has to affect a person's general health. Sure, most of the bacteria would be killed, but some could sneak past the immune system and set up house anywhere in the body. Periodontal bacteria have been found in the amniotic fluid surrounding the unborn baby. Even the placenta is not a perfect barrier.

Periodontitis is also associated with pre-eclampsia, a life-threatening disease of pregnant women, in which the blood pressure rises catastrophically. It is also linked to problems with female fertility, premature delivery, bad outcomes in pregnant diabetic women, low birth-weight babies and more.

It has also been found that infected gums can badly affect the linings of the arteries throughout the body, thereby setting the patient up for later cardiovascular disease. It is now known to be associated with abnormal ECGs, and heart disease, as well as many other diseases, e.g. diabetes, osteoporosis, some cancers (including tongue cancer) and stroke.

So treating gum disease should reduce other medical conditions, and expenses, later in life.

The Tragic Numbers

In the USA, 44 million people cannot afford medical insurance while more than 100 million do not have dental insurance. On 13 July 2007, *The Washington Post* reported on a 12-year-old child in Washington DC 'who died because of a decaying tooth that too long went untreated'.

Medicare provides all Australians with basic medical care.

But in Australia, 30 000–50 000 people are admitted to hospital every year with dental diseases so advanced that they have to be treated under general anaesthetic. In fact, the primary reason for the hospital admission of children under five years of age is a dental problem. These hospitalisations would have been preventable if the sufferers had been given affordable access to proper dental care.

Free, state-run dental services exist, but these are so underfunded that there are 650 000 people on the public waiting lists, with a waiting time of 2–6 years. If the people waiting stood 40 cm apart, the line would extend from Sydney to Canberra.

About 30% of Australians do not visit the dentist because they cannot afford it. Strangely, on the other hand, the federal government provides $500 million for dental services for those who are wealthy enough to be privately insured.

Which makes me wonder. If the mouth is part of the body and if gum disease can affect your general health, why is dental care not covered by Medicare?

Sticky-sweet Brown Drinks

If you drink colas, do not brush your teeth immediately afterwards.

The 'pH' scale – a measure of the acidity or alkalinity of a solution – runs from 14 (incredibly alkaline) to 7 (neutral) to 1 (incredibly acid, e.g., battery acid). Your teeth will be damaged by anything more acid than pH 5.5 (i.e. a pH less than 5.5).

The fizzy brown carbonated cola drinks have a pH of between 4 and 2.5. Fanta (a non-cola drink) is also very acid at pH 2.6.

Immediately after the acid has hit the enamel, it makes the surface very matte and rough, a process called dental erosion. This makes the surface temporarily very fragile.

First, if you must drink these acid drinks, do it quickly, in one hit. This gives your saliva a chance to remineralise the enamel on your teeth. The worst way to drink these acid drinks is to sip them slowly over an hour or more.

Second, if you must drink them, don't use your toothbrush straightaway because the enamel is temporarily fragile. Drink some water and swish it around to dilute the acid, wait an hour or so and then brush your teeth.

The label on the cola container will give you details of its sugar content but it won't tell you its pH. So try using a swimming pool testing kit instead.

References

Carlisle, Wendy, 'Bad teeth, bad economics', Background Briefing, 22 April 2007, http://www.abc.net.au/rn/backgroundbriefing/stories/2007/1901441.htm.

Coppa, A., et al., 'Early Neolithic tradition of dentistry', Nature, 6 April 2006, pp 755–756.

'Remember Deamont?: The Senate needs to step up on dental care for children', Washington Post, 13 July 2007.

Roan, Shari, 'Be aware of the condition of your gums', Los Angeles Times, 13 April 2007.

Roan, Shari, 'Gum disease can inflict damage elsewhere', Los Angeles Times, 16 April 2007.

Tonette, Maurizio S., et al., 'Treatment of periodontitis and endothelial function', New England Journal of Medicine, 1 March 2007, pp 911–920.

Moondance

The Moon is a major part of our life — and responsible for two-thirds of the height of the tides that wash across our beaches twice every day. In preindustrial times, it provided us with light at night and, of course, it is closely related to the female menstrual cycle.

If the Moon features so strongly in our culture, why do so many people wrongly believe that the Moon doesn't rotate? That is absolutely 100% wrong. However, if you believe that you always see exactly the same face of the Moon, you are only 9% wrong.

Celestial Mechanics 101

The Moon takes approximately 27.3 days to perform one complete orbit around the Earth — the same time that it takes for the Moon to perform one rotation on its spin axis. Yes, that's right, the Moon does spin.

By doing a simple experiment you will understand why this mythconception appears perfectly reasonable. All you will need is a rotating chair and two willing friends.

Place the rotating chair in the middle of an open space. Set down and rotate the chair slowly. (Think of yourself as the Earth.) Get a friend to walk slowly around you in a circle, always keeping their face pointing towards you. Your friend is the Moon. To you it looks quite straightforward. As the person (the Moon) walks around you (the Earth rotating in the spinning chair), you can always see the person's face. Therefore, as far as the Earth is concerned, the Moon isn't rotating.

Now get someone else to take your place in the chair. Stand yourself outside the circle that your friend, the Moon, is making around the rotating Earth. As the Moon walks around the Earth, always carefully pointing their face at the Earth, you will notice that you, the external observer, will first see the face, then one side of the head, then the back of the head, then the other side of the head, and then finally the face. (In other words, you have just observed all four sides of the person's head.) You are standing still so the person's head (the Moon) has to be rotating. So the Moon does, in fact, rotate — and in the same time that it takes to do one complete orbit around the Earth. The time for one orbit around the Earth is equal to the time for one complete rotation of the Moon.

Synchronous Rotation

It is not a coincidence that the two times are the same. Physicists call this 'synchronous rotation'.

The Earth and the Moon were created about 4.5 billion years ago – almost certainly in a giant impact, when something the size of Mars smacked into the Earth. Debris got thrown into orbit, and coalesced into the Moon. Back then, both the Moon and the Earth spun much faster than they do today. Over the eons, the Moon both slowed its spin and moved away from the Earth.

One explanation for 'synchronous rotation' relates to the so-called 'tidal force' – a force that happens when objects in space are fairly close to each other, and when the objects are fairly large. In this case, the Earth's gravity pulls quite strongly on the side of the Moon closest to the Earth – and more weakly on the side of the Moon furthest away. This tidal force actually distorts the shape of the Moon. It changes from a sphere into an ellipse pointing at the Earth. This then (via fancy physics) leads to 'synchronous rotation'.

The other explanation for 'synchronous rotation' relates to the fact that the denser parts of the Moon are on the side of the Moon facing the Earth. This makes the 'centre of gravity' of the Moon slightly offset from its 'geographical centre'. Physicists tell us that this causes a 'twisting force' or 'torque'. After only a billion years or so, this torque slowed down the rotation of the Moon, eventually leading to the Moon being gravitationally locked to the Earth.

(Think of the Moon as a barbell – with a bar joining two weights. One weight is very heavy, the other is not. The barbell is spinning quite rapidly, doing many rotations in a single orbit of the Earth. After a while, it will end up with the heavy weight pointing at the Earth, and the light weight pointing out into Space.)

In 1671, Jean Dominique Cassini discovered Iapetus, a moon of Saturn. Strangely, he could see Iapetus for only half of its orbit. The solution to this mystery came much later. First, because Iapetus is in synchronous rotation with Saturn, it always shows the same face to the planet. Second, one hemisphere is quite bright (making it easy for Cassini to see), but the other hemisphere is quite dark (making it impossible for him to see). We still do not know why Iapetus has a bright and a dark side.

Exactly the Same Face, Exactly?

The Moon does not always show the same face to the Earth. We can actually see slightly more than half (about 59%) of the Moon's surface, if you observe it over the period of a month. This effect is called 'libration', from the Latin word *libra* meaning 'a balance'.

There are three separate librations, each one allowing us to see slightly more than 50% of the Moon. Adding them together, they let us see an extra 9% of the Moon's surface.

1 – Diurnal Libration

'Diurnal' means 'relating to the day'. You can see a slightly different view of the Moon at moonrise and moonset (I bet you've never said these words before). At moonrise, you can see slightly more of the eastern side of the Moon. But at moonset you can see slightly more of the western side of the Moon. This is possible because the Moon is fairly close (384 000 km) compared to the diameter of the Earth (12 756 km).

The moon spins

The Moon takes around 27.3 days to perform one complete orbit around the Earth. And that's the same time it takes for the Moon to perform one rotation on its spin axis.

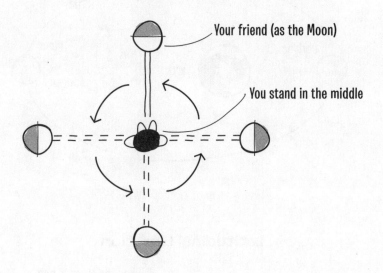

Your friend (as the Moon)

You stand in the middle

Diurnal Libration

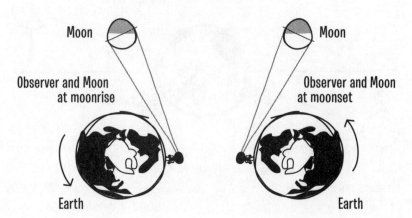

Moon

Observer and Moon at moonrise

Earth

Moon

Observer and Moon at moonset

Earth

Longitudinal Libration

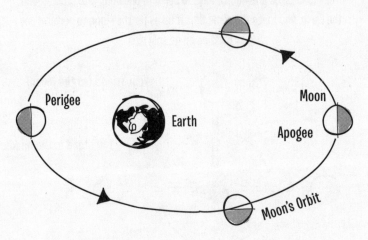

Perigee

Earth

Moon

Apogee

Moon's Orbit

Latitudinal Libration

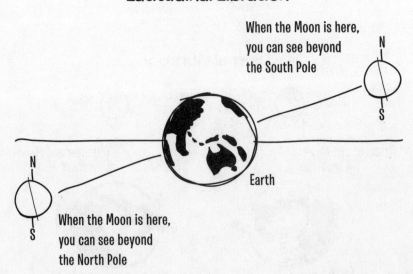

When the Moon is here, you can see beyond the South Pole

N

S

N

S

Earth

When the Moon is here, you can see beyond the North Pole

2 – Longitudinal Libration

The Moon travels in an elliptical (a squashed, circular shape) orbit around the Earth rather than in a perfect circular orbit. This means that when it is close to the Earth it moves faster and when it is further away it moves more slowly. But the Moon is always spinning at a constant rotational speed. So there is a mismatch.

When the Moon is moving slowly in its orbit we can see a little around its leading edge (top of pix), and when it is moving faster we can see a little around its trailing edge (bottom of pix).

3 – Latitudinal Libration

Finally, the Moon does not orbit around the Earth directly above the Equator. No, the Moon's orbit is tilted by 6.7° to the Equator. For this reason we can occasionally see more of the Moon's South Pole and, at other times, more of its North Pole.

So we know that the Moon does rotate and that it does not show exactly the same face to the Earth at all times. However, we still don't know why the full Moon looks so huge as it rises just above the horizon — even though you can measure it and see that it's not really bigger.

References

NASA, 'Educational Brief: Cassini Science Investigation, What is Synchronous Rotation?' EB–2001–12-003–JPL JPL 400–1018 05/02.

Phil Plait's Bad Astronomy: Misconceptions
http://www.badastronomy.com/bad/misc/moon_spin.html.

Walker, Jearl, The Flying Circus of Physics, John Wiley & Sons, 1977, pp 43, 178, 238.

CSI – Bite the Bullet

There's a veritable epidemic of 'CSI' dramas on TV — that's CSI as in Crime Scene Investigation. One common slow-motion sequence that I find really funky is the bullet barrelling through the air, pushing a shock wave in front of it. Trailing behind the bullet is a shower of fine particles (Gunshot Residue or GSR in the trade). The forensic scientist (usually a square-jawed male or a glamorous female) quickly finds these incriminating particles on the baddie, supposedly proving that the baddie has recently fired a gun. This evidence then sends the offender to the slammer — and the sleuth to small-screen success.

The science behind this scenario is horrendously inaccurate.

Gunshot Residue 101

When you pull the trigger of a gun, the 'firing pin' smashes into the blunt end of the bullet. Buried in the blunt end is the very small 'primer', which usually has three chemicals in it — the initiator (lead styphnate) that goes bang from the impact, the fuel (antimony sulphide) and the oxidiser (barium nitrate). The primer explodes when it's hit by the firing pin, igniting the much bigger main charge of explosive, usually called the 'propellant'.

The propellant (black powder or gunpowder) then turns into a huge volume of gas. This gas erupts out of the back end of the

If you can read this, you are in range.

Anatomy of a revolver

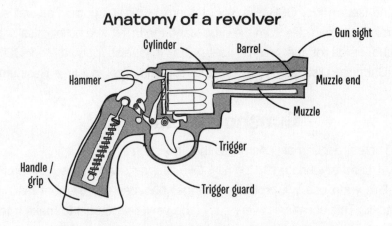

Hammer
Cylinder
Barrel
Gun sight
Muzzle end
Muzzle
Trigger
Handle / grip
Trigger guard

When you pull the trigger of a gun, the 'firing pin' smashes into the blunt end of the bullet. Buried in the blunt end is the very small 'primer'. The primer usually has three chemicals in it – the initiator (lead styphnate) that goes bang, the fuel (antimony sulphide) and the oxidiser (barium nitrate). The primer explodes when it's hit by the firing pin, igniting the much bigger main charge of explosive, called the propellant.

Anatomy of a bullet

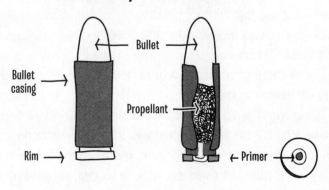

Bullet
Bullet casing
Propellant
Rim
Primer

'bullet case' or 'cartridge', in the process pushing the lead bullet in the opposite direction out of the barrel of the gun. It's a simple and very dangerous application of chemistry and Newton's Laws of Motion.

Gunshot residue is the collection of gases and solids blown out of the front of the firearm — a small amount leaks out of the back of the firearm too. GSR is a mix of chemicals in the primer, as well as miniscule particles from the propellant, the bullet, the bullet case and the inside of the firearm itself. GSR has also been called CDR (Cartridge Discharge Residue) and FDR (Firearm Discharge Residue).

Gunshot Residue 102

There are four main chemicals that 'go bang'.

Lead styphnate ($C_6Pb(NO_2)_3OH$) is a very widely used 'initiator'. This wide use will probably change, because the element lead is toxic. This chemical is very easy (and very dangerous) to make from lead oxide, styphnic acid ($C_6H_3N_3O_8$) and methanol. It crystallises into two shapes — six-sided or rectangular — and ranges in colour from yellow to brown. Lead styphnate will explode if zapped with tiny amounts of static electricity (such as from your clothing) or on slight physical impact from, say, a hammer or firing pin of a gun.

Antimony sulphide (Sb_2S_3) is the fuel in the primer. It has been used for thousands of years to make cosmetics such as rouge and eye make-up (apparently it was used to darken the eyebrows and the eyelashes). When it combines with oxygen, it burns at the right rate — not too fast and not too slow — giving off lots of heat.

Barium nitrate ($Ba(NO_3)_2$) is the 'oxidiser'. As you can see from the formula, it has six oxygen atoms.

These three chemicals have the job of generating enough heat to ignite the gunpowder.

Gunpowder is 15% charcoal, 10% sulphur and 75% potassium nitrate. When nitrogen is involved in chemical reactions, it gives off lots of energy, which is why it is used in explosives. When gunpowder burns, it gives off huge amounts of hot gas, making it ideal for use in a gun.

History of Forensics

The history behind using GSR in forensic science dates back thousands of years.

Our modern word 'forensic' comes directly from the Latin *forensis*, meaning 'forum'. In Roman times, the accused and the accuser in a criminal trial would give speeches in the forum, presenting their cases to a group of respected citizens. The orator with the better argument and/or presentation style would win.

Probably the first case of forensic science was the 'Eureka' moment of Archimedes (287–212 BC). He was trying to work out if a crown was really made of gold, not gold and lead. The legend claims that he lowered himself into a full bath, watched the water overflow, realised that the weight of water was equal to his own weight, saw in his mind how to solve the problem and joyfully ran naked through the streets shouting, 'Eureka!' ('I have found it!')

History of Firearm Identification

The earliest case of forensic firearm identification probably happened in 1784, in Lancaster, England. A certain John Toms was suspected of shooting Edward Culshaw. Back then, a piece of paper (the 'paper patch' or 'paper wad') was used to fill the gap between the lead ball and the raw gunpowder pushed into the gun. The paper patch found in the victim's head had been torn from a newspaper found in John Toms' pocket.

In 1863, during the American Civil War, the Confederate General Stonewall Jackson died on the battlefield from a bullet wound. The enemy Union soldiers used a .58 calibre Minié-ball. The bullet dug out of his body was found to be a Confederate .67 calibre ball. Jackson had been killed by 'friendly fire' from his own soldiers.

In 1900, Dr Albert Llewellyn Hall published his very prescient article 'The missile and the weapon' in the *Buffalo Medical Journal*. It dealt with the various markings that can be found on a bullet fired from a gun. However, it also discussed the various chemical residues that occur when firearms are used and how they change over time.

In 1933, Teodoro Gonzalez from the Criminal Identification Laboratory of the Mexico City Police Department invented the 'paraffin test'. It was sensitive only to the presence of the element nitrogen in the form of nitrites and nitrates. Hot paraffin wax was poured onto the back of the hands of the suspect and left to cool, allowing any GSR particles on the hand to become attached to its surface. The paraffin cast was then peeled off and treated with an acid solution of diphenylamine. This chemical turns blue in the presence of nitrates and nitrites. However, this test was not very specific — the nitrogen chemicals could have come from fertiliser, fingernail polish, striking a match or even from urine.

A more specific test was invented in 1959 by the scientists H.C. Harrison and R. Gilroy. In this test, a cotton swab moistened with various detector chemicals changes colour in the presence of antimony, barium and lead.

GSR Problem – 1

The first problem with the GSR technique is the timing of a suspect's presence in the room.

When a gun fires, there are very high temperatures and pressures in the barrel. As a result, the lead, antimony and barium turn into a gas, which then cools and condenses into tiny, invisible bumpy spheres. These spheres range in size from 0.5 to 10 microns. (To give you an idea of their size, a micron is a millionth of a metre. A human hair is about 70 microns thick.) Some of these tiny spheres are solid and can fall out of the air in just 90 seconds, while others are hollow and take up to 10 minutes to land.

Therefore, a shooter wearing gloves could fire a gun and immediately rush out of the room while tearing off the gloves, so as not to get any GSR on them. But an innocent bystander coming to investigate the incident any time within the next 10 minutes could be showered with lots of invisible GSR.

Drop, drop, drop

In 2004, Lubor Fojtá šek and Tomáš Kmječ examined the deposition rates of GSR particles by measuring how long it took for these particles to fall out of the air onto some collection plates.

They used various handguns popular in the Czech Republic, including a 9 mm Luger, a 7.65 Browning, a .38 calibre Smith and Wesson and two CZ pistols. The pistols were fired, one at a time, in a closed room measuring 5 m x 5 m x 8 m. The collection plates were 2 m from the pistol.

The time taken for the particles to land on the plates in significant numbers varied between one and a half and 10 minutes.

GSR Problem – 2

The second problem is 'cross contamination'.

GSR is like talcum powder and easily shaken or washed off the hands of the shooter. Water and a towel will remove most of it from the hands, while water, soap and a towel will remove it all. In fact, sweat is enough to wash some of it off. As you can see, it moves easily from the carrier to their local environment.

GSR is found in most American police cars, police stations and investigation rooms — places where suspects in murder cases are usually found. Therefore, it is incredibly easy for an innocent person to brush up against a contaminated surface and transfer some GSR onto themselves.

GSR Problem – 3

The third problem concerns the chemicals being tested for. Individually they are found everywhere in our modern technological society.

Lead is found in battery plates, in the materials and solder used by plumbers, in paint and even in glass. Antimony is used as a fire retardant in cotton and synthetic clothes.

And barium is found as barium sulphate in paper, as well as in automobile grease and paint.

But what about the combination of all three of these chemicals from the primer?

Although gunshots carry these chemicals, so do fireworks and the disc brake pads in cars. Therefore, most car mechanics, auto electricians and fireworks technicians will have these three chemicals on their clothing. However, the temperatures and pressures reached in car brake pads are not as high as those reached in fireworks or firearms. So while the particles from brake pads have the same chemicals as GSR (lead, antimony and barium), they have a different shape.

Specialised laboratory techniques (e.g., a scanning electron microscope equipped with an X-ray spectrometer) can identify the bumpy spheres that come from gunshots. But these specialised techniques are not in general use by police forensic laboratories around the world. And they need experienced personnel to operate the equipment and recognise the shapes of the spheres.

GSR Problem – 4

This leads to the fourth problem with the GSR technique. Different standards of analysis are used around the world.

Local law enforcement agencies often regard the presence of just two of these GSR chemicals, i.e. barium and antimony, as proof of a gunshot. And sometimes they need the third chemical, lead, to make a positive identification.

Sometimes a positive identification needs three of these microscopic particles — and sometimes it needs more, or less.

Laboratories sometimes look for just the microscopic particles containing these chemicals, but they may also use an electron microscope to examine the shapes of the particles. Therefore, in the USA, the FBI has higher standards than say, the local police in a small town. In July 2006, in Anoka, Minnesota, Judge Sharon Hall threw out gunshot residue evidence, saying that it lacked scientific backing and had no place in a courtroom.

Where does this leave 'CSI' and its copycats? Thomas Mauriello, a forensic sciences lecturer at the University of Maryland, says that about 40% of the 'science' shown on the 'CSI' TV program does not even exist. And even when the techniques do exist, the classically perfect processes of collecting and analysing the evidence shown in the programs simply do not happen in real life.

These shows are crime fiction in every sense — but on TV they still go off with a bang.

References

Bykowicz, Julie, 'Convictions tied to controversial gun-residue test', *Baltimore Sun*, 27 March 2005.

Cardinetti, Bruno, et al., 'X-ray mapping technique: a preliminary study in discriminating gunshot residue particles from aggregates of environmental occupational origin', *Forensic Science International*, 2004, Vol 143, pp 1–19.

Cole, Simon and Dioso, Rachel, 'Law and the lab: do TV shows really affect how juries vote? Let's look at the evidence', *The Wall Street Journal*, 13 May 2005.

Fojtá sek, Lubor, et al., 'Time periods of GSR particles deposition after discharge-final results', *Forensic Science International*, 2005, Vol 153, pp 32–135.

Hamby, James E., 'The History of Firearm and Toolmark Identification', *Association of Firearm and Tool Mark Examiners Journal*, Summer 1999, Vol 31, No 3, pp 266–284.

Hanes, Stephanie, 'Evidence under suspicion', *Baltimore Sun*, 23 January 2005.

Meija, Robin, 'Why we cannot rely on firearm forensics', *New Scientist*, 26 November 2005, pp 6–7.

Orrick, Dave, 'Anoka, MN judge rejects gunshot residue evidence', *Pioneer Press*, 13 July 2006.

Romolo, Francesco Saverio, 'Identification of gunshot residue: a critical review', *Forensic Science International*, 2001, Vol 119, pp 195–211.

Silver Jubilee Book Launch, Bondi Beach 2006

'Jason, Axe, I declare this rocket's going up.'

Has anyone seen Elton?

It *is* rocket science — the boys from Sydney Uni Rocket Club.

Yes, I am ready to be the first book launched — literally.

Deano, the star of 'Bondi Rocket Rescue'.

Acknowledgments

If it wasn't for the Last Minute, nothing would ever get done. In this case, the Last Minute was just a tiny while after the First Minute. So without wasting another minute, I'd like to thank the following people who made this book possible.

The Good People at the *Good Weekend* provided the initial nidus and platform for the stories in their first and shortest incarnation. Without them, these stories would never have seen the Light of Day. Dan Driscoll at the ABC both produced and edited them for radio and the web (Great Moments in Science), improving them mightily.

Alice (my elder daughter) said that she should get thanked first in The Family (she did the title of 'The Gum and the Restless'), while Caroline Pegram did most of the other titles (I'm not very good at titles). Adam Yazxhi drew all of the pictures, with a little help from his kids (Max and Zac) and my younger daughter (Lola). Mary Dobbie (my beloved wife, we're legal now) did the first (brutal) edit. Little Karl (just under 2 metres tall) also read and criticised the loose stories. At the Almost Final galleys, Max and Carmel (my parents-in-law) pored over the work, finding errors that had somehow snuck through.

My editors, Janice Godwin and Lydia Papandrea, edited energetically. My agents, Lesley McFadzean and Sophie Hamley, agented bravely, while Alison Urquhart and Shona Martyn published magnificently. And Judi Rowe and Jill Donald professionally produced it all.

Finally, let me thank the scientists/doctors/engineers/linguists etc, who actually did the original research that I merely wrote about. I especially thank the researchers who read the stories I wrote about their work, and reviewed and improved them. They include Jenni Brand-Miller and Tim Crowe ('Celery with Negative Kilojoules'), Andy Johnston ('She Smell Seasmells (Ozone) …'), Bruce Edelman ('The Gum and the Restless'), Thomas Mauriello ('CSI? Bite the Bullet') and Anthony Woodbury ('They Snow This, They Snow That …').

Other Dr Karl titles

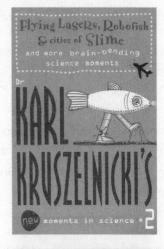

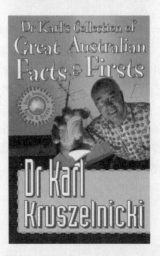

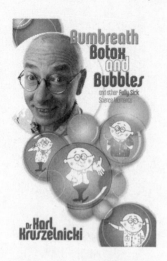

Bumbreath
Botox
and
Bubbles
and other Fully Sick
Science Moments

Dr **Karl**
Kruszelnicki

Great
MYTH
CONCEPTIONS

DEPARTMENT OF MYTHCONCEPTIONS

Dr. K. Kruszelnicki
Field Agent: 5275 4 7738

Dr **Karl**
KRUSZELNICKI

Dis Information

and other
Wikkid
myths

DR
KARL
KRUSZELNICKI

WARNING

IT
AIN'T
NECESSARILY
SO...bro

DR KARL
KRUSZELNICKI